**CABI CONCISE**

FOOD NANOTECHNOLOGY SERIES

Nanotechnology can be applied to a range of food production processes to enhance flavour, quality, safety, nutritional value and shelf-life. The use of innovative nanomaterial technologies (like nanosensors, nano-encapsulation, nano-emulsions and nanoparticles) can thus provide better preservation and packaging as well as the safe delivery of micronutrients and bioactive compounds.

This series looks at the latest breakthroughs in food nanoscience and how these can be used to improve the sustainability, safety and cost of food production. It thus provides a range of instructive guides on the recent trends, advances and obstacles in food nanotechnology focused on the benefits and limitations of this rapidly evolving science.

**Series Editor**

Neha Srivastava,
Department of Chemical Engineering and Technology, Indian Institute of Technology (BHU), Varanasi, India
Email: sri.neha10may@gmail.com

## Titles Published

Nanotechnology in Functional Foods and Nutraceuticals
*Gurudatta Singh, Monika Yadav and Pardeep Singh*

Nanotechnology for Smart Food Processing: Sustainable Nanoscale Solutions
*Gurudatta Singh, Monika Yadav and Pardeep Singh*

## Titles in Preparation

Nanosensors for Food Safety
*Dan Bahadur Pal and Ashish Kapoor*

# Nanotechnology for Smart Food Processing

## Sustainable Nanoscale Solutions

**Gurudatta Singh**
*Institute of Environment and Sustainable Development (IESD), Banaras Hindu University (BHU), Varanasi, India-221005*

**Monika Yadav**
*Department of Earth and Environmental Science, Parul Institute of Applied Sciences, Parul University, Vadodara, 391760, Gujarat, India*

**Pardeep Singh**
*Department of Environment Studies, Pannalal Girdharlal Dayanand Anglo-Vedic College, University of Delhi, New Delhi, India-110065*

CABI

**CABI is a trading name of CAB International**

CABI
Nosworthy Way
Wallingford
Oxfordshire OX10 8DE
UK

Tel: +44 (0)1491 832111
E-mail: info@cabi.org
Website: www.cabi.org

CABI
200 Portland Street
Boston
MA 02114
USA

T: +1 (617)682-9015
E-mail: cabi-nao@cabi.org



A catalogue record for this book is available from the British Library, London, UK.

ISBN-13: 9781836994695 (hardback)
9781836994701 (paperback)
9781836994718 (ePDF)
9781836994725 (ePub)

DOI: 10.1079/9781836994725.0000

Commissioning Editor: Jamie Lee
Editorial Assistant: Theresa Regueira
Production Editor: Theresa Regueira

Typeset by Exeter Premedia Services Pvt Ltd, Chennai, India
Printed in the USA

# Contents

## Dedication

*This book is dedicated to the innovators, researchers, and visionaries tirelessly working to transform the food industry into a sustainable and efficient domain through nanotechnology. To the farmers, food scientists, and technologists who strive for excellence in ensuring food security and quality for future generations, your commitment inspires progress. To the researchers, scientists, and innovators who relentlessly explore the frontiers of science, pushing the boundaries of what is possible, your dedication and passion inspire this endeavor. I also dedicate this work to the students and future scientists who will continue exploring and expanding nanotechnology's possibilities in food processing. May this book serve as a stepping stone in your journey toward discovering innovative solutions that will shape the future of the food industry. Lastly, to the readers, may this book spark curiosity, foster innovation, and contribute to a healthier, more sustainable world for all.*

# Preface

The global food industry stands at a critical juncture, facing the dual challenges of ensuring sustainability and meeting the demands of a rapidly growing population. With increasing pressure to minimize resource consumption, reduce food waste, and enhance product quality, innovative solutions are more essential than ever. Nanotechnology for Smart Food Processing: Sustainable Solutions emerges as a timely response to these challenges, exploring the transformative potential of nanotechnology in revolutionizing the food processing sector. This book was conceived to bridge the gap between cutting-edge nanotechnological advancements and their practical applications in food systems. It delves into how nanoscale innovations can enhance efficiency, improve food safety, and extend shelf life while aligning with sustainability goals. This work addresses the pressing need for scalable and environmentally responsible technologies in food production and processing, from encapsulation techniques for bioactive compounds to nanofiltration systems for waste management. Written with a multidisciplinary audience in mind, this book is designed for researchers, food technologists, environmental scientists, policymakers, and industry professionals eager to harness the benefits of nanotechnology. It also aims to inspire students and early-career scientists by showcasing this field's possibilities for innovation and sustainable development. The chapters are thoughtfully organized to guide readers through the foundational principles of nanotechnology and its wide-ranging applications. Case studies, recent breakthroughs, and practical insights are included to stimulate interest and encourage further exploration. Whether seeking a comprehensive introduction to nanotechnology or specific applications to enhance your work, this book is valuable.

# Acknowledgements

The journey of creating "*Nanotechnology for Smart Food Processing Techniques*" has been supported by many individuals and institutions to whom we owe a deep gratitude. First and foremost, we extend my heartfelt thanks to my colleague and teacher for providing the support and resources necessary to undertake this project. The encouragement and flexibility offered by the organization have been instrumental in bringing this book to fruition. We do not have any funding agency for this proposal. Special thanks go to the academic institutions and colleagues collaborating on this project and sharing their insights and expertise. Their contributions have enriched the content and broadened the scope of this work.

# 1

# Current Trends and Market Potential of Nano-enabled Food Technologies

## 1.1 Introduction

Nano-enabled food technologies refer to the application of nanoscience and nanomaterials (NMs) in food processing, packaging, preservation, and nutrient delivery (Dasgupta *et al.*, 2018; He *et al.*, 2019). These technologies offer transformative advantages, including improved food safety, enhanced nutrient bioavailability, controlled release of food additives, and active packaging systems that extend shelf life (Dasgupta *et al.*, 2018; Awlqadr *et al.*, 2025). As consumer demand grows for healthier, safer, and more sustainable food products, nano-enabled innovations emerge as key drivers in the food industry. Recent advances include the development of nanosensors for pathogen detection, nanocarriers for targeted nutrient delivery, and nanocomposites for innovative packaging (de Sousa *et al.*, 2023; Ahmad *et al.*, 2025). Food manufacturers are increasingly adopting these innovations, supported by investments in research and development (R&D) and favorable regulatory frameworks in regions such as North America, Europe, and the Asia-Pacific. However, adopting these technologies also faces challenges, including regulatory concerns, toxicity assessments, and consumer acceptance (Dasgupta *et al.*, 2018; Ahmad *et al.*, 2025).

Despite these hurdles, the global market for nano-enabled food technologies is projected to grow rapidly in the coming years, driven by technological innovation and the need for sustainable food systems. Integrating nano-enabled technologies in food processing, packaging, and distribution reshapes food production and consumption (Finglas *et al.*, 2014; Nile *et al.*, 2020). These innovations utilize nanoscale materials and techniques to address persistent issues in the food industry, including spoilage, contamination, and the loss of nutritional quality. This chapter explores emerging trends, commercial advancements, and the growing market prospects of nano-enabled technologies in the food sector, emphasizing global developments and future outlooks (Duran and Marcato, 2013; Finglas *et al.*, 2014; Nile *et al.*, 2020). Nanotechnology is widely applied in everyday life and is changing society.

Nanotechnology has emerged as a transformative force across multiple industries, including agriculture and food systems. Its integration into the food sector began gaining structured attention in 2003 with the roadmap released by the United States Department of Agriculture (DOA, 2003; He *et al.*, 2019). Since then, rapid advancements have driven its application in enhancing food quality, safety, and processing efficiency. The global nanotechnology market has experienced substantial growth, reaching approximately US$79–91 billion in 2023–2024 and projected to expand significantly in the coming decade (Nile *et al.*, 2020; Ahmad *et al.*, 2025). Earlier projections estimated its economic impact at nearly US$3 trillion, with employment opportunities for around 6 million people worldwide. Within this broader domain, the food nanotechnology sector has evolved into a multibillion-dollar market, estimated at around US$14 billion in 2024, with expected steady growth driven by increasing demand for advanced food systems (Ahmad *et al.*, 2025; Awlqadr *et al.*, 2025; The Business Research Company, 2025).

A major application of nanotechnology in the food industry is nano-enabled packaging, which enhances barrier properties, extends shelf life, and improves food preservation. This segment alone accounts for a significant market share, with values in the tens of billions of US dollars and projected growth rates in the low-to-mid teen percentages (Nile *et al.*, 2020). Innovations such as nano-encapsulation for targeted nutrient delivery, nanosensors for food safety monitoring, and smart packaging systems are contributing to the rapid expansion of this field. The growth of nanotechnology in the food sector is driven by several key factors, including the need for improved food safety and traceability, longer shelf life, and the development of functional foods with enhanced nutritional value (Fig. 1.1; Bryksa and Yadav, 2012; de Sousa *et al.*, 2023; Ahmad *et al.*, 2025). Cross-sector advancements, particularly in health care, electronics, and materials science, have further accelerated innovation in food-related applications (Table 1.1). Despite its promising potential, the commercialization of nanotechnology in food systems faces several challenges. Regulatory concerns, safety assessments, and consumer acceptance remain critical issues that influence its adoption (Bryksa and Yadav, 2012; de Sousa *et al.*, 2023; Ahmad *et al.*, 2025). Most current applications involve indirect contact with food, such as packaging and coatings, rather than direct incorporation into consumable products. Only a few NMs, such as titanium dioxide ($TiO_2$) and iron oxide, have been used directly in food as additives like pigments and colorants (He *et al.*, 2019). Nanotechnology continues to revolutionize the food industry by enabling the development of advanced materials and innovative solutions. However, ensuring safety, regulatory compliance, and public trust will be essential for its sustained growth and widespread acceptance in the future (Magnuson *et al.*, 2011; Singh *et al.*, 2022).

Over recent decades, nanotechnology has emerged as a transformative technology in the food sector, operating at the nanoscale (1–100 nm) to

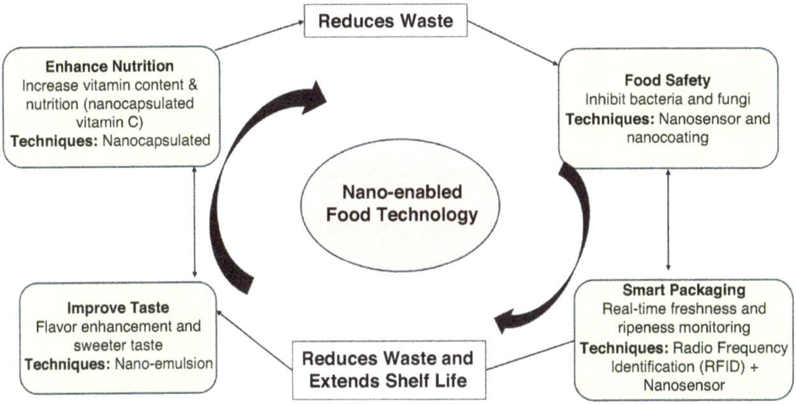


**Fig. 1.1.** Nano-enabled food technologies.

manipulate atoms and molecules for creating materials with novel properties (Singh *et al.*, 2017, 2022). These NMs exhibit unique physicochemical characteristics such as enhanced solubility, strength, diffusivity, optical and magnetic properties primarily due to their high surface-to-volume ratio. Unlike bulk materials, their behavior enables improved functionality in food applications. Nanotechnology has driven a new industrial revolution, attracted significant global investment and offered wide-ranging applications in food, agriculture and medicine (Singh *et al.*, 2017; He *et al.*, 2019).

Rising consumer concerns about food quality and health have driven the adoption of nanotechnology to enhance food without compromising nutritional value. Nanoparticle-based materials are increasingly used due to their stability under high temperature and pressure and generally low toxicity (Dasgupta *et al.*, 2018; Yadav *et al.*, 2025). Nanotechnology provides solutions across food manufacturing, processing, and packaging, significantly improving food quality, safety, and functionality. Applications in the food sector are broadly categorized into nanostructured ingredients and nanosensing (Nile *et al.*, 2020). Nanostructured ingredients involve engineered nanoparticles (NPs) that enhance the delivery, stability, and bioavailability of nutrients, flavors, and bioactive compounds (Table 1.2; Fig. 1.2). Techniques such as nano-encapsulation protect sensitive components like vitamins, omega-3 fatty acids, and probiotics, enabling controlled release and improved absorption. Nanocarriers such as liposomes and polymeric NPs are widely used in functional foods for targeted nutrient delivery (Nile *et al.*, 2020; Chudasama and Goyary, 2024; Chen *et al.*, 2025). Food nanosensing focuses on nanosensors that detect contaminants, pathogens, and quality changes in real time. These sensors, often integrated into packaging, can monitor freshness, spoilage, and harmful substances like pesticides or allergens (Khan *et al.*, 2025). For example, gold or silver NP-based sensors enable rapid detection of microbial contamination through visible changes.

**Table 1.1.** Nanomaterials in the food industry: direct and indirect use.

| Nanomaterial | Direct use (in food) | Indirect use (packaging/processing) | Applications | References |
|---|---|---|---|---|
| Nanosilver (AgNPs) | Not approved for direct ingestion in most countries | Used in antimicrobial food packaging, coatings, and storage containers | Antimicrobial packaging, shelf-life extension | Carbone *et al.*, 2016; Trotta *et al.*, 2023 |
| Nanotitanium dioxide ($TiO_2$) | Sometimes used as a food additive (E171), but under scrutiny | Used in ultraviolet (UV)-protective food packaging | Whitening agent, UV protection in packaging | Blaznik *et al.*, 2021 |
| Nanosilica ($SiO_2$) | Used as an anti-caking agent (E551) | Packaging for moisture regulation | Flow aid, moisture absorption | Singh *et al.*, 2023a |
| Zinc oxide nanoparticles (ZnO NPs) | Not approved for direct consumption | Antimicrobial agent in packaging and coatings | Antibacterial films and surface treatments | Peters *et al.*, 2012 |
| Chitosan nanoparticles | Used in nano-encapsulation of bioactives | Used in edible films and coatings | Delivery of nutraceuticals, antimicrobial coating | Muñoz-Tebar *et al.*, 2023 |
| Lipid-based nanoparticles (nano-emulsions, SLNs, NLCs) | Directly used for the delivery of flavors, nutrients, and drugs | Primarily used in the food matrix, not the packaging | Improved solubility and bioavailability of compounds | Ashfaq *et al.*, 2023 |
| Carbon nanotubes (CNTs) | Not approved for use in food | Used in food sensors (e.g. contamination detection) | Biosensors, spoilage, and pesticide detection | Singh *et al.*, 2022; Kumar *et al.*, 2023; Sharma *et al.*, 2023b |
| Nanoclays | Not used directly in food | Packaging barrier improvements | Enhancing the mechanical and barrier properties of packaging | Deshmukh *et al.*, 2023 |

| | | | | |
|---|---|---|---|---|
| Iron oxide nanoparticles ($Fe_3O_4$ NPs) | Used for iron fortification in food and supplements | Also used in the magnetic separation of food contaminants | Iron supplementation, food safety | Kumari and Chauhan, 2022; Sarode and Mahajan, 2024 |
| Dendrimers | Potential for drug/nutrient delivery | Not widely used in packaging | Targeted delivery of bioactives | |
| Nanocalcium carbonate ($CaCO_3$) | Used in mineral-fortified food and beverages | Used in food-contact materials | Calcium enrichment, mechanical reinforcement | Singh *et al.*, 2022; Zhao *et al.*, 2022 |
| Gold nanoparticles (AuNPs) | Investigated for food colorants and diagnostics | Biosensing and pathogen detection in food | Diagnostics, colorimetric sensors | |
| Nanostarch | Used in food processing and digestion control | Used in packaging and edible films | Fat replacer, texture modification | Adeyeye *et al.*, 2023 |
| Nano encapsulated essential oils | Nano-encapsulation protects Essential oils (EOs) from degradation and allows for their controlled release, significantly improving their effectiveness as natural food preservatives and extending shelf life | Nanomaterials can create "smart" packaging that indicates the freshness or spoilage of food through color changes or other visual signals, providing real-time information to consumers and preventing waste | Controlled release of natural antimicrobials improves safety and shelf life | Liao *et al.*, 2021 |
| Graphene oxide nanosheets | The detection and removal of patulin from apple products were highlighted due to the contaminant risk since the apple industry can create a larger range of food production | Graphene is being examined to see how it could offer coatings to reduce the carbon footprint in the packaging supply chain | High tensile strength and antimicrobial effects in food packaging | Hashim *et al.*, 2022 |

*Continued*

**Table 1.1.** Continued

| Nanomaterial | Direct use (in food) | Indirect use (packaging/ processing) | Applications | References |
|---|---|---|---|---|
| Biopolymer nanocomposites | Use nanoscale fillers, such as nanoparticles, to reinforce natural polymers, dramatically improving their mechanical properties and barrier functions | Smart packaging | Enabled pH and freshness indicators for real-time monitoring | Dasgupta *et al.*, 2025; Singh *et al.*, 2025b |
| Nutrient delivery | Nano-encapsulation of vitamins, minerals, antioxidants | Nanocarriers ensuring controlled release during storage | Various bodily functions are maintained, and food-derived bioactive components provide health benefits | Pugazhendhi *et al.*, 2025 |
| Flavor enhancement | Nanomicelles for flavor solubility and stability | Nanostructured films controlling flavor migration in packaging | improved food texture, taste, and storage properties, alongside benefits such as reduced spoilage, better barrier properties, and the development of active and intelligent packaging systems | Chudasama and Goyary, 2024 |
| Edible coatings | Nanochitosan films for microbial resistance | Nanosilica reinforcing edible packaging | They are applied to the fruits and vegetables to prolong shelf life and enhance storage quality | Xing *et al.*, 2019; Yadav *et al.*, 2025b |
| Color and appearance | Nanotitanium dioxide for whitening, opacity | Nanostructured coatings preventing color loss in packaging | Applied as active packaging films to block UV and visible light that degrade natural food pigments (like carotenoids and anthocyanins). Enhances opacity in beverages and sauces, making them visually appealing | Piryaei and Azimi, 2024 |

Nanotechnology enhances food processing, preservation, and safety while improving nutritional outcomes across the supply chain (Table 1.2; Ezhilarasi *et al.*, 2013; Yadav *et al.*, 2025).

## 1.2 Current Status of Food Nanotechnology

Food nanotechnology has infiltrated many aspects of customer products, such as food packaging, additives, and preservation. The recognition of this novel technology has advanced food processing and storage to ensure food safety (Table 1.2). Many conventional chemicals added as food additives or packaging materials have also been found to exist partially at the nanometer scale. For example, food-grade $TiO_2$ NPs have now been found up to approximately 40% in the nanometer range (Nile *et al.*, 2020). Although NMs like $TiO_2$ NPs are generally recognized as low toxic at ambient conditions, long-term exposure to such NMs may cause adverse damage (Dorier *et al.*, 2017; He *et al.*, 2019). The application of novel food nanotechnology and the presence of nanoscale chemicals have also drawn public attention to potential risks. In this section, we carefully review the current progress on the application of food nanotechnology (Gautam and Avasthi, 2020). The United States Food and Drug Administration (US FDA) and the European Commission (EC) are the primary sources for legislation and regulation on food nanotechnology (Nile *et al.*, 2020; Shafiq *et al.*, 2020). Some authorizations made by the US FDA and EC are based on the risk assessment of a substance's conventional particle size; therefore, the authority may require a case-by-case basis for engineered NPs.

### 1.2.1 Nanotechnology in food processing

The nanostructured food ingredients are being developed to improve taste, texture, and consistency. Nanotechnology is increasing the shelf life of different kinds of food materials and also helps reduce the extent of food wastage due to microbial infestation (Pradhan *et al.*, 2015; Shafiq *et al.*, 2020; Chhabra *et al.*, 2025). Nanocarriers are being utilized as delivery systems to carry food additives in food products without disturbing their basic morphology. Currently, several types of nanocarriers are in use or under commercialization.

#### *1.2.1.1 Lipid-based nanocarriers*

- **Nano-emulsions**: Used for encapsulating vitamins, flavors, antioxidants, and preservatives. Example: clear beverages enriched with omega-3.
- **Solid Lipid Nanoparticles (SLNs)**: For controlled release of lipophilic additives (like carotenoids).
- **Nanostructured Lipid Carriers (NLCs):** Improved stability for sensitive nutraceuticals (e.g. coenzyme Q10, vitamin D).

**Table 1.2.** Current trends and market potential of nano-enabled food technologies.

| Application area | Nanotechnology type | Key innovations /products | Current trends | Market potential | References |
|---|---|---|---|---|---|
| Food packaging | Nanoclay, nanosilver, nanocellulose | Antimicrobial packaging, oxygen scavengers and moisture barriers | Innovative packaging, biodegradable nanocomposites | Projected to reach US$25.4 billion by 2030 | Dasgupta *et al.*, 2025 |
| Food preservation | Nano-emulsions, nano-encapsulation | Controlled release of preservatives, improved shelf life | Use of nano-emulsions for natural preservatives (e.g. essential oils) | Increasing demand for clean-label and preservative-free foods | Agnihotri *et al.*, 2025 |
| Nutraceutical delivery | Liposomes, solid lipid nanoparticles (SLNs) | Targeted delivery of vitamins, omega-3, and probiotics | Nano-encapsulation of bioactives for enhanced bioavailability | The functional food market is expected to reach US$530 billion by 2027, with nanotech playing a key role | Chhabra *et al.*, 2025 |
| Food additives and ingredients | Nanosilicon dioxide, nanotitanium dioxide | Anti-caking agents, color enhancers | Regulatory scrutiny is increasing in the EU and the US; there is a trend toward safer, generally recognized as safe (GRAS)-certified nano-additives | Moderate market share; focus on safety and regulation | Chaudhary *et al.*, 2008; Pandhi *et al.*, 2023 |
| Food quality and safety sensing | Nanosensors, quantum dots, nanowires | Detection of pathogens, toxins, temperature/time indicators | Integration into Internet of Things (IoT)-enabled packaging systems | Smart food sensor market expected to exceed US$8 billion by 2030 | Shafiq *et al.*, 2020 |
| Flavor and texture enhancement | Nano-encapsulated flavors, nano-emulsifiers | Enhanced flavor release, better solubility of hydrophobic compounds | Nanoflavor carriers are being adopted in beverages and confectionery | Growth driven by consumer demand for premium sensory experiences | Cunha *et al.*, 2024 |
| Food fortification | Nanocalcium, nano-iron, nanozinc | Improved solubility, stability, and absorption of micronutrients | Fortification of staple foods in developing countries | Essential in tackling malnutrition; strong support from public health organizations | Kumari and Chauhan, 2022 |
| Functional beverages | Nanoliposomes, nanocoacervates | Stable delivery of polyphenols, caffeine, and plant extracts | High market growth in wellness drinks and sports beverages | The beverage sector is adopting nanocarriers for increased functionality | Pandey *et al.*, 2025a |

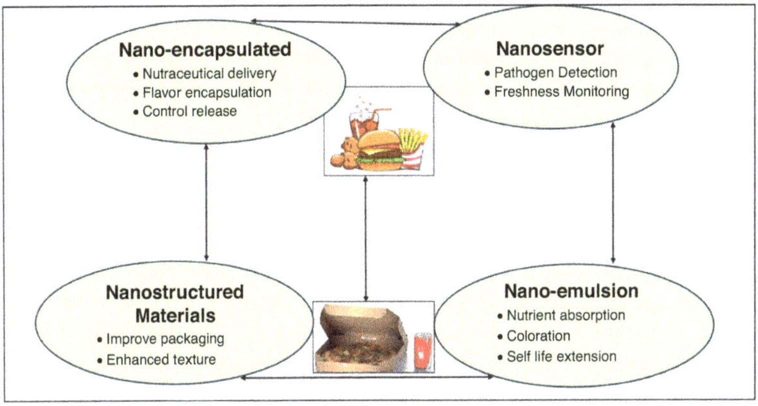


**Fig. 1.2.** Current trends and market potential of nano-enabled food technologies.

### *1.2.1.2 Polymeric nanocarriers*

- **Biopolymer nanoparticles (chitosan, alginate, zein, starch, casein):** Used to encapsulate flavors, probiotics, essential oils, and natural preservatives.
- **Dendrimers (limited in food so far):** Used experimentally for encapsulating antioxidants.

### *1.2.1.3 Protein-based nanocarriers*

- **Casein micelles**: Natural nanocarriers present in milk; now engineered to deliver calcium, vitamin D, and hydrophobic nutraceuticals.
- **Soy protein and whey protein nanoparticles**: Applied for encapsulating curcumin, polyphenols, and omega-3.

### *1.2.1.4 Polysaccharide-based nanocarriers*

- **Cyclodextrins**: Widely used in food products to encapsulate flavors and vitamins and remove undesired tastes or odors.
- **Starch nanoparticles**: Used as carriers for flavors and active compounds.

Particle size may directly affect the delivery of any bioactive compound to various sites within the body, as it was noticed that in some cell lines, only submicron NPs can be absorbed efficiently, but not the larger-sized microparticles (Singh *et al.*, 2017). An ideal delivery system should fulfill the following criteria: (i) targeted release of the active compound at the precise site of action, (ii) controlled release at a specific time and rate, and (iii) prolonged stability of the active compound under storage conditions. Nanotechnology offers innovative solutions to meet these requirements through the development of encapsulation systems, nano-emulsions, biopolymer-based matrices, simple nanodispersions, and association colloids (Gautam and Avasthi, 2020; Shafiq *et al.*, 2020). These nano-enabled delivery systems enhance bioavailability, protect bioactive compounds, and ensure efficient, site-specific delivery. In addition, nanopolymers are increasingly being explored as alternatives to conventional materials in food packaging due to their superior barrier, mechanical and antimicrobial properties (Yadav *et al.*, 2025). Nanosensors provide advanced tools for detecting contaminants, mycotoxins, and pathogenic microorganisms in food, contributing significantly to food safety and quality monitoring (Paliwal *et al.*, 2025).

Nanoparticles have better properties for encapsulation and release efficiency than traditional ones (Pandhi *et al.*, 2023). Nano-encapsulations mask odors or tastes, control interactions of active ingredients with the food matrix, control the release of the active agents, ensure availability at a target time and specific rate, and protect them from moisture, heat (He *et al.*, 2019; Amir *et al.*, 2025), chemical, or biological degradation during processing, storage and utilization, and also exhibit compatibility with other compounds in the system (He *et al.*, 2019; Pandhi *et al.*, 2023).

| Nanocoatings for packaging and preservation | | |
|---|---|---|
| Nanomaterial | Function | Example in packaging/ food contact |
| Silver nanoparticles (AgNPs) | Antimicrobial, prolong shelf life | Fresh produce, meat, and dairy packaging |
| Zinc oxide (ZnO) nanoparticles | UV-blocking, antimicrobial | Beverage bottles, plastic films |
| Nanoclay composites | Barrier to gases ($O_2$, $CO_2$) | Snack wrappers, frozen food packs |
| Chitosan nanoparticles | Biodegradable antimicrobial coating | Fruits, cheese, bakery items |
| Nanosensors for food safety | | |
| Nanomaterial | Function | Example |
| Gold nanoparticles (AuNPs) | Detect pathogens or toxins | Rapid detection in milk or juice |
| Carbon nanotubes (CNTs) | Detect chemical contamination | Smart packaging labels |
| Quantum dots | Fluorescent detection of spoilage | Biosensors for seafood and meat |

These delivery systems can penetrate deeply into tissues due to their small size, enabling efficient delivery of active compounds to target sites in the body (He *et al.*, 2019). Various synthetic and natural polymer-based encapsulation delivery systems have been developed to improve the bioavailability and preservation of active food components. The importance of nanotechnology in food processing can be evaluated by considering its role in the improvement of food products in terms of (i) food texture, (ii) food appearance, (iii) food taste, (iv) nutritional value of the food, and (v) food shelf life.

These delivery systems can penetrate deeply into tissues due to their small size, enabling efficient delivery of active compounds to target sites in the body (He *et al.*, 2019). Various synthetic and natural polymer-based encapsulation delivery systems have been developed to improve the bioavailability and preservation of active food components. The importance of nanotechnology in food processing can be evaluated by considering its role in the improvement of food products in terms of (i) food texture, (ii) food appearance, (iii) food taste, (iv) nutritional value of the food, and (v) food shelf life.

#### *1.2.1.5 Nutritional value for nano-enabled foods*

Most bioactive compounds, such as lipids, proteins, carbohydrates, and vitamins, are sensitive to a highly acidic environment and the enzyme activity of the stomach and duodenum (Singh *et al.*, 2017; Rezagholizade-Shirvan *et al.*, 2024). Encapsulation of these bioactive compounds enables them to resist such adverse conditions and assimilate readily in food products, which is quite hard to achieve in non-capsulated form due to the low water-solubility of these bioactive compounds (Agnihotri, 2025). Nanoparticle-based tiny edible capsules to improve the delivery of medicines, vitamins, or fragile micronutrients in daily foods are being created to provide significant health benefits (Singh *et al.*, 2017). The nanocomposite, nano-emulsification, and nanostructuration are different techniques that have been applied to encapsulate substances in miniature forms to more effectively deliver nutrients like protein and antioxidants for precisely targeted nutritional and health benefits (Shafiq *et al.*, 2020). Polymeric NPs are found to be suitable for the encapsulation of bioactive compounds (e.g. flavonoids and vitamins) to protect and transport bioactive compounds to target functions (Rojas *et al.*, 2024).

Nanotechnology has revolutionized various sectors, and its application in food science has significantly enhanced food products' nutritional value, bioavailability and safety. Integrating nanotechnology in food systems primarily involves nano-encapsulation, nanosensors and nanocarriers, which improve nutrient delivery, protect sensitive bioactives and enable controlled release mechanisms (Yousefian Rad *et al.*, 2020). This advanced approach has opened new possibilities for developing functional foods, fortifying staple products with essential micronutrients, and preventing nutritional deficiencies, particularly in vulnerable populations (Rezagholizade-Shirvan *et al.*, 2024). One of the key areas where nanotechnology contributes to nutritional value is through nano-encapsulation of vitamins, minerals and bioactive compounds. Many nutrients, such as omega-3 fatty acids, vitamin D, curcumin, and polyphenols, are unstable or have poor solubility in aqueous systems (Fig. 1.3 and Table 1.3). Encapsulation into nanostructures like liposomes, polymeric NPs, or nanoemulsions protects them from degradation due to environmental factors such as heat, light, and oxygen (Yadav *et al.*, 2025). It enhances their bioavailability

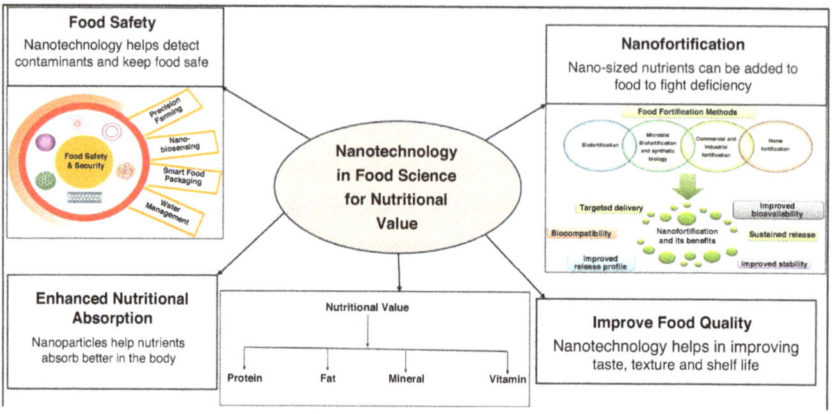


**Fig. 1.3.** Nanotechnology in food science for nutritional value.

by facilitating easier absorption in the gastrointestinal tract. For example, curcumin, a polyphenolic compound with antioxidant and anti-inflammatory properties, shows low absorption in its natural form (Rezagholizade-Shirvan *et al.*, 2024).

| Nano-additives for color, flavor, and texture | | |
|---|---|---|
| Nanomaterial | Function | Example in Food Products |
| Nanotitanium dioxide ($TiO_2$) | Whitening, opacity enhancement | Used in confectionery, chewing gum, and sauces |
| Nanosilver (AgNPs) | Antimicrobial properties, extending shelf life | Coatings on meat packaging, dairy products |
| Nano-encapsulated flavors (liposomes, polymeric NPs) | Controlled release of flavors | Beverages, snacks, and bakery products |
| Nanoclays | Improve texture, prevent caking | Spices, powdered food mixes |
| Nano-encapsulation for nutrient delivery | | |
| Nanomaterial/carrier | Function | Example in food products |
| Liposomes | Encapsulate vitamins, omega-3 fatty acids | Fortified dairy, beverages |
| Solid lipid nanoparticles (SLNs) | Controlled release of bioactives | Nutraceutical drinks, functional foods |
| Nano-emulsions | Enhance solubility and bioavailability | Vitamin D-fortified beverages, salad dressings |
| Polymeric nanoparticles (chitosan, poly lactic acid (PLA)) | Stabilize sensitive compounds | Probiotics, antioxidant-enriched foods |

Nanotechnology plays a pivotal role in enhancing the nutritional profile of staple foods such as rice, wheat flour, and milk by facilitating the precise and uniform incorporation of essential micronutrients such as iron, zinc, and vitamins without altering the food's sensory attributes such as taste, color, or texture (Singh *et al.*, 2017; Ardhi *et al.*, 2025). For example, nano-encapsulated iron has been successfully employed to fortify foods without imparting the unpleasant metallic flavor commonly associated with conventional iron supplements (Akanda *et al.*, 2023; Rezagholizade-Shirvan *et al.*, 2024). Nanotechnology-based packaging systems contribute indirectly to nutrition preservation by extending shelf life, minimizing nutrient degradation, and preventing microbial contamination (He *et al.*, 2019; Ardhi *et al.*, 2025). Advanced nanosensors integrated into innovative packaging can track parameters such as temperature, gas exchange, and microbial activity in real time, ensuring food quality and safety throughout storage and distribution (Silvestre *et al.*, 2011; Yadav *et al.*, 2025).

#### *1.2.1.6 Texture, taste, and appearance of food*

Nanotechnology provides a range of options to improve food quality and enhance food taste. Nano-encapsulation techniques have been used broadly to improve the flavor release and retention and to deliver culinary balance (Nakagawa, 2014; Cheng *et al.*, 2025). Zhang *et al.* (2014) used nano-encapsulation for the highly reactive and unstable plant pigment anthocyanins, which have various biological activities. Encapsulating cyanidin-3-O-glucoside (C3G) molecules within the inner cavity of the recombinant soybean seed H-2 subunit ferritin (rH-2) improved the thermal stability and photostability. This design and fabrication of multifunctional nanocarriers for bioactive molecule protection and delivery (Rezagholizade-Shirvan *et al.*, 2024; Pandey *et al.*, 2025). Rutin is a common dietary flavonoid with great importance in pharmacological activities. However, due to poor solubility, its application in the food industry is limited (Fig. 1.4; Biswas *et al.*, 2022; Pandey *et al.*, 2025). The ferritin nanocages encapsulation enhanced the solubility, thermal, and UV radiation stability of ferritin-trapped rutin as compared to free rutin (Biswas *et al.*, 2022). The use of nano-emulsions to deliver lipid-soluble bioactive compounds is becoming increasingly popular since they can be produced using natural food ingredients, using easy production methods, and may be designed to enhance water dispersion and bioavailability (Nakagawa, 2014; Ozturk *et al.*, 2015; Pandey *et al.*, 2025). Compared to larger particles, which generally release encapsulated compounds more slowly and over extended periods, NPs provide a promising means of improving the bioavailability of nutraceutical compounds due to their subcellular size, leading to a higher drug bioavailability (Table 1.4; Rezagholizade-Shirvan *et al.*, 2024; Ardhi *et al.*, 2025). Many metallic oxides, such as $TiO_2$ and silicon dioxide ($SiO_2$), have conventionally been used as color or flow agents in food items (Ottaway, 2009; Ozturk *et al.*, 2015; Pandey *et al.*, 2025). $SiO_2$ NMs are also one of

**Table 1.3.** Nutrients enhanced using nanotechnology and their benefits.

| Nutrient/ compound | Typical nanoformulation | Food/product examples | Key benefits of nanotechnology | References |
|---|---|---|---|---|
| Iron (Fe) | Nanostructured iron salts, iron nanoparticles, FePO4-NPs, encapsulated iron in lipids/polymers | Fortified rice, cereal, powdered beverages, composite staples | Higher intestinal absorption (bioavailability) vs bulk iron; reduced sensory (taste/color) changes and lower pro-oxidant effects in food matrices. Helpful in treating iron deficiency with lower doses | Baumgartner *et al.*, 2022 |
| Vitamin D (cholecalciferol/ ergocalciferol) | Nano-emulsions, liposomes, solid-lipid nanoparticles (SLN) | Fortified milk, yogurt drinks, beverages, capsules | Improved water-dispersibility and intestinal uptake; greater stability in processing/storage; more uniform fortification of beverages | Miah *et al.*, 2025 |
| Omega-3 fatty acids (EPA/DHA) | Nano-emulsions, micro/nano-encapsulation in proteins or lipids | Fortified dairy, spreads, beverages and infant formula | Reduced oxidation (off-flavors), improved physical stability, higher bioaccessibility and absorption, protected delivery through the GI tract | Pandey *et al.*, 2025 |
| Vitamin B-group (example: riboflavin / $B_2$, $B_{12}$) | Polymer nanoparticles, chitosan-coated poly (lactic-co-glycolic acid) (PLGA), NPs, liposomes | Nutraceutical tablets, fortified powdered mixes | Enhanced stability against light/ processing; controlled release and improved cellular uptake for people with absorption issues | Sathiensathaporn *et al.*, 2025 |
| Vitamin C (ascorbic acid) | Nano-encapsulation in polymer or lipid carriers, nano-emulsions | Fortified juices, powdered supplements | Protection from oxidation during processing/storage; controlled release and better retention of activity in foods | Maurya *et al.*, 2023 |

| | | | | |
|---|---|---|---|---|
| Fat-soluble vitamins (A, E, carotenoids) | Nano-emulsions, nanostructured lipid carriers, solid-lipid nanoparticles | Fortified oils, margarines, beverages and functional foods | Increased water-compatibility (enables incorporation into aqueous products), higher bioaccessibility/bioavailability and protection from degradation (heat, light, oxygen) | Ardhi *et al.*, 2025 |
| Carotenoids (β-carotene, lycopene, lutein) | Nano-emulsions, nanocapsules, protein-based nanoparticles | Beverages, spreads, infant foods, supplements | Improved solubility and GI uptake, better color stability and masking of off-tastes; controlled release in gut to increase conversion/absorption | Sridhar *et al.*, 2021 |
| Curcumin (polyphenol/ antioxidant) | Lipid nanoparticles, nano-emulsions, polymeric nanocarriers | Functional beverages, supplements, nutraceuticals | Dramatic increase in apparent bioavailability vs free curcumin; improved solubility/stability and targeted intestinal delivery | El-Saadony *et al.*, 2025 |
| Probiotics (live beneficial microbes) | Nano/micro-encapsulation, layer-by-layer coatings, polymer nanoparticles | Yogurts, synbiotic capsules, functional foods | Increased survival through gastric acid and bile, better viability during storage and delivery to the intestine → higher effective dose reaching the colon | Cheng *et al.*, 2025 |
| Folic acid (folate) | Nanocarriers (polymeric/ lipid), nano-encapsulation | Fortified flour, beverages, prenatal supplements | Improved stability (against heat/ light), reduced interaction with food matrix and more consistent bioavailability in fortified staples | Ashraf *et al.*, 2025 |

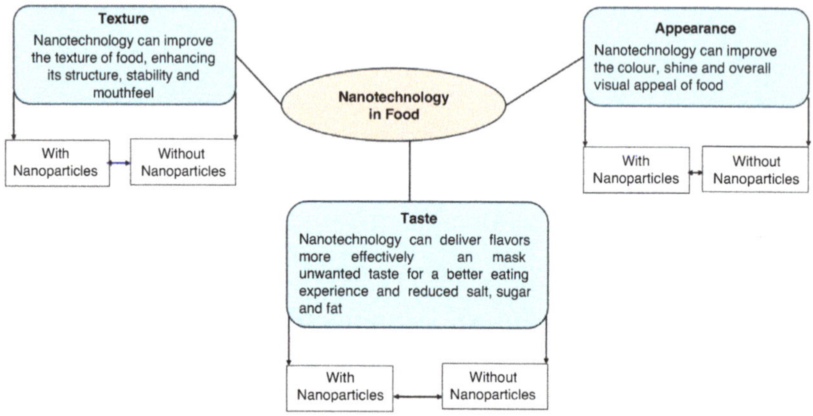


**Fig. 1.4.** Texture, taste, and appearance of food in nanotechnology.

the most used food NMs as carriers of fragrances or flavors in food products (Nakagawa, 2014; Ozturk *et al.*, 2015; Pandey *et al.*, 2025).

One of the most significant impacts of nanotechnology is in improving food texture. The mouthfeel of products such as dairy items, meat analogs, and beverages can be modified using nano-emulsions and nanostructured ingredients (He *et al.*, 2019; Chudasama and Goyary, 2024). Nano-emulsions with droplet sizes in the 20–200 nm range provide a smooth and creamy texture, often without increasing fat content (Sharma *et al.*, 2023; Yadav *et al.*, 2025). This is especially valuable in low-fat or fat-free products, where traditional fat removal produces undesirable textures (Rezagholizade-Shirvan *et al.*, 2024). Nanocellulose, another nanostructured material, is used as a rheology modifier to enhance the gel-like texture in sauces and dressings without synthetic additives (Aliabbasi and Emam-Djomeh, 2024; Ardhi *et al.*, 2025).

In terms of taste, nanotechnology can enhance flavor delivery and masking mechanisms. Flavors and bioactives encapsulated in NPs, such as liposomes or cyclodextrins, can be engineered to release only under certain conditions (e.g. changes in pH or temperature) (Chudasama and Goyary, 2024; Pandey *et al.*, 2025). This targeted release ensures that flavors are experienced at the right moment, intensifying the sensory perception (Sharma *et al.*, 2023). Moreover, bitter or undesirable tastes from functional ingredients, such as plant-based proteins or nutraceuticals, can be effectively masked. Nanoscale salt or sugar crystals can also reduce overall intake while maintaining taste intensity, addressing health concerns without compromising sensory quality (Sharma *et al.*, 2021a; Chudasama and Goyary, 2024).

**Table 1.4.** Applications of nanotechnology in enhancing the sensory properties of food.

| Sensory attribute | Nanotechnology application | Nanomaterial used | Mechanism of enhancement | Example/product | Reference |
|---|---|---|---|---|---|
| Color | Nano-encapsulation of natural pigments | Nano-emulsions, liposomes | Protects pigments from degradation, improves stability and dispersion in the food matrix. | Nano-encapsulated beta-carotene in beverages | Chaudhary *et al.*, 2022 |
| Color | Enhanced color vibrancy in packaging | Silver nanoparticles (AgNPs) | Interaction with light improves appearance and shelf display | Color-improved juice bottles | Rana *et al.*, 2025 |
| Flavor | Controlled release of flavor compounds | Cyclodextrin inclusion complexes, lipid-based nanocarriers | Protects and gradually releases volatile flavor molecules | Encapsulated vanilla or mint flavors | Vasile *et al.*, 2025 |
| Flavor | Reduction of off-flavors in fortified foods | Nano-iron, nanozinc carriers | Avoids direct interaction with flavor compounds | Iron-fortified flour with no metallic aftertaste | |
| Aroma | Enhanced aroma retention and stability | Solid-lipid nanoparticles (SLNs), nanogels | Prevents evaporation and oxidation of aroma molecules | Aromatic oils in bakery and confectionery | Rajendran and Vadakkepushpakath, 2024 |
| Texture | Modified mouthfeel and creaminess | Nanocellulose, nanostarch, nano-emulsions | Alters rheology, improves smoothness and viscosity | Fat mimetics in low-fat yogurt | Rajendran and Vadakkepushpakath, 2024 |
| Texture | Stabilization of foams and emulsions | Pickering emulsions using nanoparticles | Enhances the structural stability of food products | Whipped toppings, mayonnaise | |
| Taste | Masking unpleasant tastes (e.g. bitterness) | Polymer-based nanocapsules, liposomes | Encapsulation delays or modifies taste perception | Bitter vitamins in gummies | Coupland and Hayes, 2014 |
| Multi-sensory appeal | Smart delivery of sensory enhancers | Responsive nanocarriers | Triggered release under pH, temperature, or enzymatic conditions | Functional chewing gum or bright candies | Karimi *et al.*, 2016; Rajendran and Vadakkepushpakath, 2024 |

## 1.3 Concept of Nano-enabled Food Technologies

Nano-enabled food technologies represent a cutting-edge convergence of nanoscience and food science, where NMs and nanoscale processes are employed to improve food quality, safety, functionality, and sustainability (Singh *et al.*, 2017). The application of nanotechnology in the food sector encompasses the entire food value chain from agriculture and food processing to packaging, distribution, and consumption (Sharma *et al.*, 2024a; Pandey *et al.*, 2025b). At the nanoscale, materials exhibit unique physicochemical properties such as increased surface area, reactivity, and enhanced interaction with biological systems, which are leveraged to develop more intelligent, more effective food systems (Rezagholizade-Shirvan *et al.*, 2024). One of the significant applications of nanotechnology in food processing is the development of nano-encapsulation systems. These systems deliver bioactive compounds such as vitamins, antioxidants, flavors, and antimicrobials with greater stability and targeted release (Biswas *et al.*, 2022; Pandey *et al.*, 2025b). Nanocarriers, including liposomes, nano-emulsions, dendrimers, and polymeric NPs, protect sensitive nutrients from degradation and enhance their bioavailability (Chudasama and Goyary, 2024; Muthu *et al.*, 2025). For example, omega-3 fatty acids, prone to oxidation, can be encapsulated in nanoliposomes to improve shelf life and maintain nutritional quality (Muthu *et al.*, 2025; Pandey *et al.*, 2025a).

Nanosensors are another significant innovation in nano-enabled food technologies. These sensors are used for real-time food quality monitoring, detecting spoilage, pathogens, or toxic contaminants at very low concentrations (Awlqadr *et al.*, 2025; Muthu *et al.*, 2025). Nanosensors based on carbon nanotubes (CNTs), quantum dots, or metal oxide NPs can be integrated into packaging materials to develop intelligent packaging systems (Rezagholizade-Shirvan *et al.*, 2024). Such systems can signal consumers about product freshness, temperature history, or tampering events, enhancing food safety and traceability (Ghaani *et al.*, 2024; Pandey *et al.*, 2025b). In food packaging, nanocomposites are employed to enhance mechanical strength, barrier properties, and antimicrobial functionality. Incorporating NPs such as nanosilver, nanozinc oxide, or nanoclay into biopolymer matrices leads to packaging materials that can extend the shelf life of perishable products (Rezagholizade-Shirvan *et al.*, 2024).

## 1.4 Conclusion

Incorporating nanotechnology into the food industry has profoundly reshaped traditional food processing, packaging, preservation, and safety methods. Nano-enabled technologies in the food sector represent a multidisciplinary innovation, merging concepts from materials science, biotechnology, and food engineering to tackle existing challenges related to food quality, shelf-life extension, efficient nutrient delivery, and contamination control. This chapter explores the latest developments in this domain, emphasizing applications such as nanostructured materials in innovative packaging, nanosensors for real-time spoilage detection, and nano-encapsulation techniques that facilitate targeted and sustained delivery of bioactive compounds. These advancements collectively demonstrate nanotechnology's transformative potential in enhancing food products functional, sensory, and nutritional profiles. From a market standpoint, the global nano-enabled food industry is experiencing rapid growth, fueled by increasing consumer demand for safer, minimally processed, and nutritionally enriched food products. Major industry players are accelerating the integration of nanotechnology through the development of food-grade NMs, affordable manufacturing technologies and, in some regions, favorable regulatory policies. Recent market insights reveal significant growth in sectors such as functional foods, nutraceuticals and active packaging, underscoring the strong commercial viability of nanofood technologies. Growing consumer awareness of the health advantages of nanoformulated foods and rising R&D investments by food technology firms continue to expand the market landscape.

Despite these advancements, several critical barriers hinder the widespread deployment of nano-enabled technologies in the food sector. Concerns persist regarding the potential toxicity of engineered NMs, their behavior within biological systems and the absence of standardized protocols for assessing associated risks. Questions around NMs' environmental fate, mobility, and potential bioaccumulation necessitate further scientific inquiry. Regulatory authorities, including the FDA, European Food Safety Authority (EFSA) and Food and Agriculture Organization (FAO)/WHO, have issued preliminary guidelines; however, globally harmonized standards are still under development. To build consumer confidence and ensure responsible innovation, transparent risk communication, comprehensive labeling, and active public engagement are imperative. The future of nanofood technologies appears promising, especially with the rise of sustainable approaches such as green synthesis methods, biodegradable NMs, and artificial intelligence (AI)-powered optimization of production processes.

# 2

# Role of Nanotechnology in Reducing Food Waste: Sustainability Prospects

## 2.1 Introduction

Food waste presents significant challenges for the global food system, contributing to environmental degradation, economic loss, and social inequities. Approximately one-third of all food produced for human consumption is globally projected to be wasted (FAO, 2011, 2019). According to a report by the FAO of the United Nations, the world wasted an estimated 1.05 billion tons of food in 2022 across retail, food service, and household sectors, equating to 132 kg per person per year, of which 79 kg per capita came from households (FAO, 2022). This wastage ultimately causes exhaustion of natural resources. The wastage occurs at several supply chain stages, from production and processing to distribution and consumption (Chauhan *et al.*, 2021). The implications of such waste are profound, as it not only exacerbates food insecurity but also contributes to greenhouse gas (GHG) emissions, with wasted food ending up in landfills, generating substantial amounts of methane. The Intergovernmental Panel on Climate Change (IPCC) identifies methane as at least 28 times as effective as carbon dioxide in trapping heat over a century. According to FAO, food waste accounts for approximately 8–10% of global GHG emissions, encompassing both loss and waste, and occupies nearly 30% of the world's agricultural land (Environment UN, 2024).

The environmental consequences of food waste reveal significant economic ramifications. The United Nations estimates that the direct economic impact of food waste amounts to approximately US$750 billion each year, excluding seafood and fish. When seafood and fish are included, this figure rises to nearly US$1 trillion (Sharma *et al.*, 2023). The issue of food loss is further complicated by the water resources utilized in food production. A substantial portion of the world's freshwater is dedicated to agriculture, meaning that wasted food also represents wasted water. Each year, around 250 $km^3$ of freshwater are lost due to food that is ultimately discarded. In fact, about a quarter of all freshwaters on Earth are consumed in the production of food that will never be eaten (Sharma *et al.*, 2023). By 2050, it is projected that nearly 4 billion people will be affected by these issues (https://www.theworldcounts.com/challenges/people-and-poverty/hunger-and-obesity/food-waste-statistics).

Such economic losses highlight the need for targeted interventions aimed at reducing food waste across the supply chain. Therefore, valorizing food waste presents both economic and environmental opportunities that can help alleviate the challenges linked to its traditional disposal methods (Dhiman *et al.*, 2025). Research on the valorization of food waste has gained momentum due to its potential to mitigate environmental and economic challenges. However, several research gaps persist that hinder the effective implementation of valorization strategies (Sharma *et al.*, 2023). One significant gap is the limited understanding of the economic and environmental challenges associated with various valorization methods. Another area needing attention is the diverse composition of food waste, which varies widely depending on the source and type of food waste. This variability complicates the selection of appropriate valorization technologies (Biswas *et al.*, 2022). Most existing research focuses on specific types of waste, such as fruit and vegetable residues, while neglecting other categories like dairy or meat by-products that may require different processing approaches. Moreover, there is a lack of standardized methodologies for evaluating the efficacy of valorization techniques across different waste types (Fig. 2.1).

Figure 2.2 highlights the global share of various food waste generated across the globe. The implications of such waste are profound, as it not only exacerbates food insecurity but also contributes to GHG emissions, with wasted food ending up in landfills, generating substantial amounts of methane (Emmanuel *et al.*, 2025). The IPCC identifies methane as at least 28

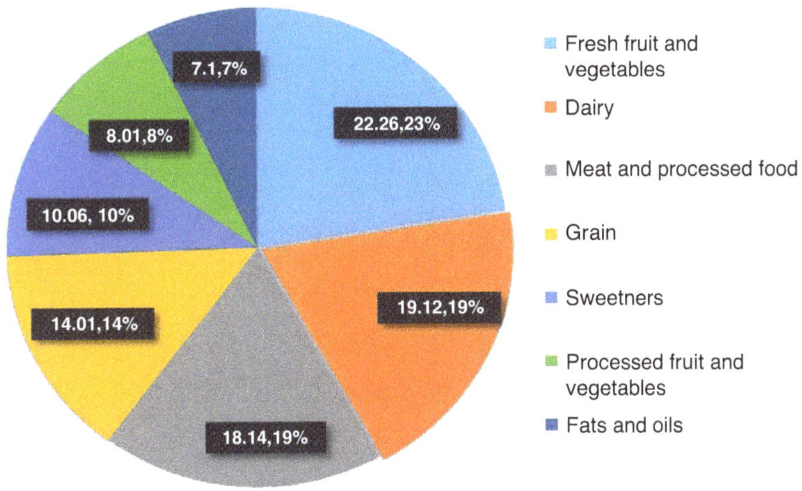


**Fig. 2.1.** Global share of various food waste sources.

times as effective as carbon dioxide in trapping heat over a century (Kilemile *et al.*, 2025). According to FAO, food waste accounts for approximately 8–10% of global GHG emissions, encompassing both loss and waste, and occupies nearly 30% of the world's agricultural land (UNEP, 2024). Nanotechnology has emerged as a promising avenue to address global sustainability challenges, particularly in reducing food waste throughout the supply chain (Fig. 2.3). Nearly 1.3 billion tons of food waste annually exacerbate hunger and

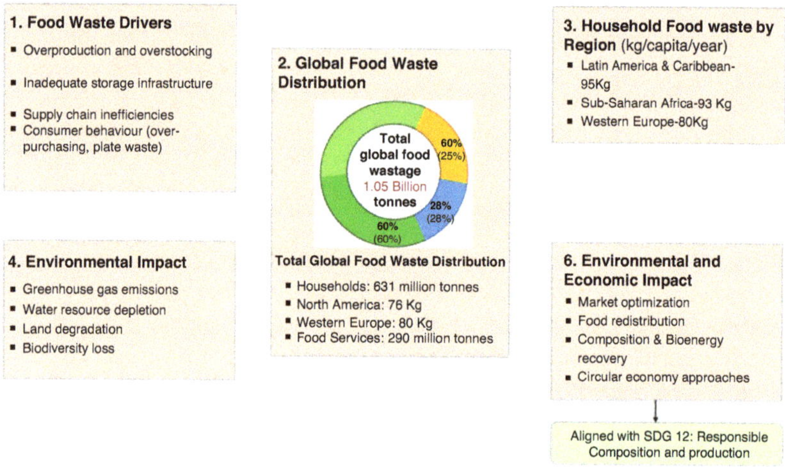


**Fig. 2.2.** Sources of food waste in the world scenario.

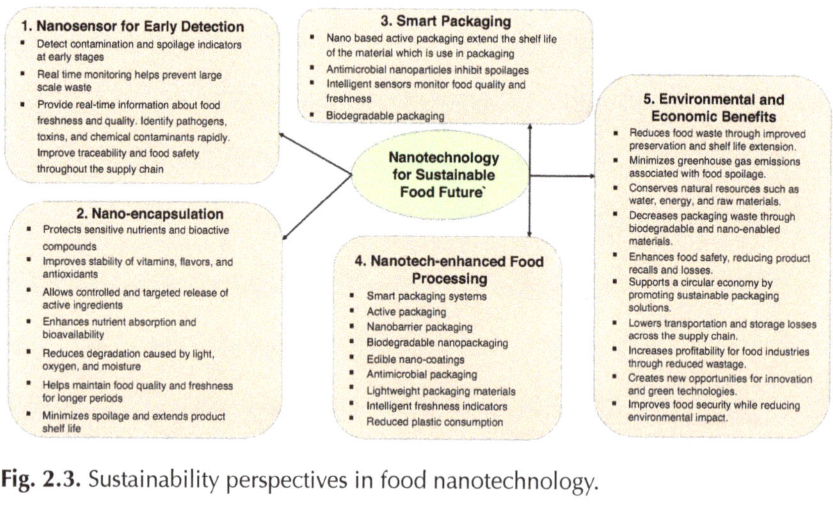


**Fig. 2.3.** Sustainability perspectives in food nanotechnology.

contribute significantly to GHG emissions and resource depletion (Sharma *et al.*, 2023; UNEP, 2024).

Nanotechnology also supports intelligent packaging systems through nanosensors capable of detecting temperature changes, gas emissions (e.g. ethylene, $CO_2$, $NH_3$), and microbial metabolites (FAO, 2022). These nanosensors can be embedded into packaging films or labels, enabling continuous, real-time monitoring of food quality, freshness, and potential spoilage (Sharma *et al.*, 2023). Such smart packaging systems facilitate traceability, extend shelf life, and improve logistics management, especially in cold chain systems (Biswas *et al.*, 2022). This, in turn, empowers stakeholders, including producers, retailers, and consumers, to make informed decisions and reduce unnecessary discards across the food supply chain (Emmanuel *et al.*, 2025). Moreover, edible nanocoatings made from chitosan, starch, and cellulose composites integrated with NPs (e.g. silver, zinc oxide, or titanium dioxide) offer biodegradable and nontoxic protective layers (Biswas *et al.*, 2022; Sharma *et al.*, 2023). These coatings can inhibit microbial growth, regulate moisture exchange, and act as oxygen barriers without altering the sensory attributes of food. For instance, nanosilver has demonstrated potent antimicrobial properties, while nanoclay-based films improve mechanical strength and barrier properties (Liao *et al.*, 2021). These nanocoatings contribute to reduced dependency on synthetic plastic packaging, thereby addressing environmental concerns related to plastic waste (Ghosh *et al.*, 2025). Advanced developments are also exploring responsive packaging materials that change color in response to spoilage or contamination, offering visual indicators to consumers (Biswas *et al.*, 2022).

Nanostructured lipid carriers and SLNs are further being studied for the controlled release of bioactive compounds (e.g. antioxidants, antimicrobials), enhancing food safety and quality during storage (Sharma *et al.*, 2023). However, despite these benefits, deploying nanotechnology in food systems necessitates rigorous assessment of environmental and health impacts. The potential migration of NPs into food products, their long-term toxicity, bioaccumulation in human tissues, and impacts on gut microbiota are ongoing areas of concern (Sharma *et al.*, 2023). Regulatory frameworks remain fragmented and inconsistent across countries, creating uncertainty for industry stakeholders (Dam *et al.*, 2025). There is a pressing need for standardized testing protocols, transparent labeling, and consumer education to support safe commercialization (Ankiel *et al.*, 2025). Minimizing spoilage and postharvest losses enables these systems to significantly reduce GHG emissions, optimize resource utilization, and enhance the overall sustainability of food systems. With ongoing advancements in research and the continuous evolution of safety and regulatory frameworks, nanotechnology demonstrates substantial potential in supporting sustainability objectives in food preservation and packaging (Table 2.1; Fig. 2.4; Sharma *et al.*, 2023; Ankiel *et al.*, 2025).

**Table 2.1.** Nanotechnology applications in reducing food waste and their sustainability implications.

| Nanotechnology application | Mechanism/ function | Food waste reduction benefit | Sustainability implications | References |
|---|---|---|---|---|
| Nanopackaging (e.g. nanoclay, nanosilver, ZnO) | Enhances barrier properties against oxygen, moisture, and microbial ingress | Extends shelf-life, prevents spoilage during storage and transport | Reduces packaging material needs, improves food security | Sharma *et al.*, 2023; Gupta *et al.*, 2024b |
| Nanosensors for spoilage detection | Detects gases (e.g. ethylene, $CO_2$, $NH_3$) or microbial metabolites | Real-time monitoring of freshness, enabling timely consumption | Minimizes premature discards and over-packaging | FAO, 2022; Awlqadr *et al.*, 2025 |
| Edible nanocoatings (e.g. chitosan, cellulose, starch) | Forms biodegradable protective films on produce | Delays ripening, inhibits microbial growth | Reduces plastic use, biodegradable waste generation | Emmanuel *et al.*, 2025 |
| Nano-encapsulation of preservatives/ antioxidants | Controlled release of active compounds like essential oils | Maintains food quality, slows oxidation | Minimizes chemical use and enhances efficiency | Das *et al.*, 2021; Liao *et al.*, 2021 |
| Nano-emulsions for antimicrobial sprays | Uniform dispersion of antimicrobials on food surfaces | Prevents contamination and postharvest losses | Eco-friendly food preservation with reduced residues | Das *et al.*, 2021; Biswas *et al.*, 2022 |
| Smart nanolabels (e.g. time-temperature indicators) | Provides real-time data on food conditions | Reduces reliance on expiration dates and food wastage | Promotes circular economy practices | Muthu *et al.*, 2025 |
| Nanocarriers for nutrient fortification | Delivers vitamins and minerals effectively | Enhances value of surplus or rejected produce | Supports nutritional sustainability and reduces loss | Elemike *et al.*, 2019 |
| Nanozymes for contamination sensing | Mimics enzyme activity to detect toxins/pathogens | Ensures food safety and reduces batch discards | Prevents resource-intensive recalls | Liao *et al.*, 2021 |

*Continued*

**Table 2.1.** Continued

| Nanotechnology application | Mechanism/ function | Food waste reduction benefit | Sustainability implications | References |
|---|---|---|---|---|
| Nanopackaging | Barrier properties, antimicrobial, oxygen scavenging | Nanosilver embedded biodegradable films extending shelf life of meat | SDG 2 (Zero Hunger), SDG 12 (Responsible Consumption and Production) | Sharma *et al.*, 2025 |
| Nanocoatings | Edible protective layers preventing moisture loss and microbial growth | Chitosan–ZnO nanocomposite coatings on fruits and vegetables | SDG 3 (Good Health and Well-being), SDG 12 | Sapna *et al.*, 2025 |

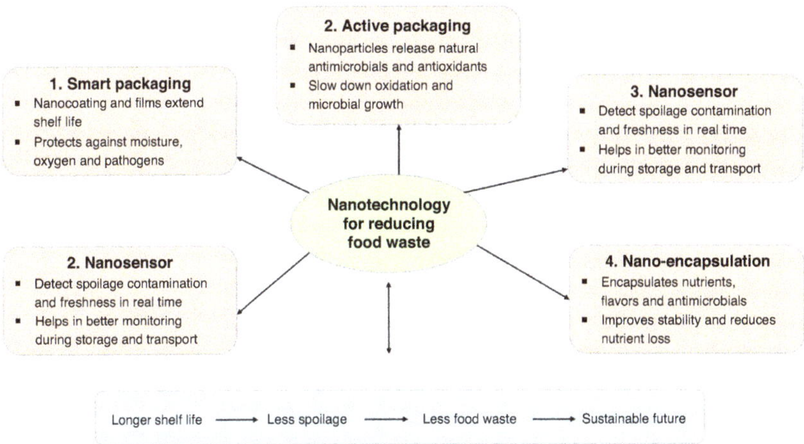


**Fig. 2.4.** Role of nanotechnology in reducing food waste.

## 2.2 Analysis of the Relationship Between Food Waste and Sustainable Development Goals (SDGs), Including SDG 2, SDG 12, SDG 13, and SDG 14

Food waste refers to the reduction of food intended for human consumption along the supply chain due to loss, damage, disposal, or diversion to other uses (FAO, 2019). A major cause of food waste is poor coordination among stakeholders across the food supply chain (Elemike *et al.*, 2019). Waste occurs at all stages, from production and processing to retail and final consumption. In the early stages, inadequate infrastructure, inefficient postharvest handling, and poor processing techniques contribute significantly to food loss (Manzoor *et al.*, 2024). At later stages, especially in retail, hospitality, and households, consumer behavior and overpurchasing are major contributors (Sharma *et al.*, 2023; Manzoor *et al.*, 2024). Nearly one-third of edible food produced globally is wasted, creating substantial economic, environmental, and social consequences (Zambrano-Zaragoza *et al.*, 2020; Das *et al.*, 2021). This paradox, often described as "scarcity within abundance," highlights the coexistence of high food waste and persistent food insecurity (Galli *et al.*, 2019).

Food waste generation is further driven by population growth, urbanization, industrialization, and globalization, which alter consumption patterns and increase purchasing power (Liao *et al.*, 2021). Managing food waste is challenging due to its diverse composition, variable generation sources, and increasing volume. Globally, food waste has emerged as a major sustainability concern, closely linked to food security, resource conservation, and climate change. It directly relates to several Sustainable Development Goals (SDGs), including Zero Hunger (SDG 2), Responsible Consumption and Production (SDG 12), Climate Action (SDG 13), and Life Below Water (SDG 14) (Table 2.2; Fig. 2.5). According to the United Nations, approximately 1.3 billion tons of food are lost annually, while millions remain undernourished (Prasad *et al.*, 2022; Emmanuel *et al.*, 2025). At the same time, overconsumption has led to rising obesity levels, particularly in developed and rapidly developing nations (Muthu *et al.*, 2025; Sharma *et al.*, 2025). Food waste also contributes significantly to environmental degradation. The food sector accounts for about 22% of global GHG emissions and 30% of total energy consumption (Sapna *et al.*, 2025). Changing dietary patterns toward resource-intensive foods further increases emissions and environmental pressures. Therefore, adopting sustainable consumption practices, improving supply chain efficiency, and implementing circular economy strategies are essential to reduce food waste and support global sustainability goals. Researchers propose implementing modern thermal and nonthermal processing methods to enhance food safety and sustainability, utilizing secure, clean, and energy-efficient technology (Mohanta *et al.*, 2024). This approach promotes clever land use, efficient factory production, and reduced GHG emissions, aligning with SDG 13 to combat climate change. SDG 14 (Awlqadr *et al.*, 2025) focuses on conserving marine ecosystems and sustainable resource use, aiming to manage fish stocks for maximum sustainable yield. Achieving this goal requires addressing illegal and overfishing practices, alongside reducing fish losses and waste through improved governance and management (Sapna *et al.*, 2025). Lowering the food waste of fish products supports sustainable resource utilization and aligns with SDG 14 objectives. By eliminating harmful subsidies and discouraging discarding, reducing food waste emerges as a viable strategy to advance SDG 14 (Awlqadr *et al.*, 2025).

**Table 2.2.** Enhanced food preservation and packaging – A sustainability perspective.

| Preservation/ packaging technology | Mechanism | Sustainability impact | Advantages | References |
|---|---|---|---|---|
| Edible coatings with nanomaterials | Coatings made from biopolymers (e.g. chitosan, alginate) embedded with nanoparticles (e.g. ZnO, $TiO_2$) | Reduces plastic waste and prolongs shelf life | Biodegradable, safe, and enhances shelf life without altering taste | Karnwal *et al.*, 2025a; Negi and Yadav, 2025 |
| Modified atmosphere packaging (MAP) | Altering $O_2$, $CO_2$, and $N_2$ concentrations inside packaging | Extends freshness, minimizes food loss | Delays microbial growth and enzymatic activity | Boz *et al.*, 2018 |
| Active packaging (e.g. $O_2$ scavengers, moisture absorbers) | Incorporates active agents to interact with the internal environment | Prevents spoilage, reduces chemical preservatives | Improves food safety and quality | Sahoo *et al.*, 2021; Yadav *et al.*, 2025 |
| Intelligent packaging (nanosensors) | Monitors real-time conditions like temperature, gas, and spoilage indicators | Minimizes unnecessary discarding by providing freshness indicators | Enhances traceability and reduces food waste | Dobrucka, 2016; Fuertes *et al.*, 2016a |
| Biodegradable polymers (PLA, PHA, starch-based films) | Derived from renewable biomass and decomposes under composting | Replaces conventional petroleum-based plastics | Environmentally friendly and renewable | Dallaev *et al.*, 2025 |
| Cold plasma preservation | Ionized gas disrupts microbial cell membranes | Energy-efficient non-thermal method | Maintains nutritional and sensory qualities | Harikrishna *et al.*, 2023 |
| High pressure processing (HPP) | Uses extreme pressure to inactivate pathogens | Extends shelf life without additives | Retains food quality, lowers chemical use | Lisboa *et al.*, 2024 |
| Ultraviolet (UV-C) irradiation | UV-C damages the DNA of spoilage organisms | Low energy requirement and no chemical residues | Effective surface sterilization | Tchonkouang *et al.*, 2023 |
| Recyclable and reusable packaging | Packaging designed for multiple cycles or full recyclability | Reduces single-use material consumption | Promotes circular economy | Zhu *et al.*, 2022 |

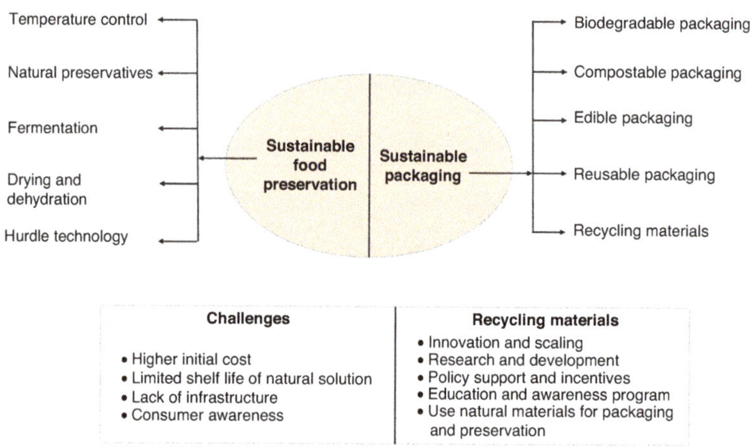


**Fig. 2.5.** Sustainable approaches for food preservation and packaging.

## 2.3 Improved Food Preservation and Packaging

Nanotechnology offers innovative solutions for antimicrobial food packaging, helping to extend shelf life, minimize spoilage, and reduce waste. By incorporating antimicrobial nanocomposites into packaging materials, bacterial growth can be inhibited, thereby preserving food quality for longer periods (Shaaban and Farouk, 2022). Integrating nanosensors into packaging enables real-time monitoring of freshness, supporting better inventory control and reducing the likelihood of food disposal as products near their expiration dates (Brody *et al.*, 2008; Sharma *et al.*, 2025). Such advancements in food preservation and packaging are vital for reducing food loss, ensuring global food security, and promoting environmental sustainability (Chen *et al.*, 2024). Modern preservation technologies focus on extending shelf life, maintaining nutritional value, and preserving sensory qualities while limiting the use of chemical preservatives (Sharma *et al.*, 2023). At the same time, packaging serves not only to protect food from physical, chemical, and biological damage but also as a key medium for communication and marketing (Cheng *et al.*, 2024). From a sustainability standpoint, innovations in both preservation and packaging must align with ecological and economic objectives, aiming to lower environmental impacts and improve resource efficiency throughout the food supply chain (Negi and Yadav, 2025; Yadav *et al.*, 2025). Emerging preservation techniques such as high-pressure processing, pulsed electric fields, cold plasma, and UV irradiation have shown promise in inactivating spoilage microorganisms without the need for high temperatures, thus retaining the quality of heat-sensitive food products (Biswas *et al.*, 2022). These nonthermal methods offer energy-efficient alternatives to conventional thermal processing. Additionally, bio-based preservatives derived from natural sources such as essential oils, bacteriocins, and enzymes have been explored as safer and more environmentally friendly alternatives to synthetic additives (Biswas *et al.*, 2022). Simultaneously, smart and sustainable packaging solutions have been developed to complement advanced preservation techniques (Singh *et al.*, 2016). Biodegradable and compostable materials, such as those derived from starch, cellulose, chitosan, and polylactic acid (PLA), are gaining traction as alternatives to traditional petroleum-based plastics (Biswas *et al.*, 2022; Ghosh *et al.*, 2025).

## 2.4 Existing Challenges of Nanomaterials in Sustainable Food Preservation and Packaging

Nanotechnology has revolutionized food packaging by enhancing mechanical, thermal, and barrier properties while maintaining cost-effectiveness and sustainability. It improves durability, temperature resistance, and enables controlled release of antimicrobial and antioxidant agents, thereby extending product shelf life (Pattnaik *et al.*, 2024). Nanocomposites are widely used in food packaging and coatings for fruits and vegetables, where they delay ripening, reduce spoilage, and support food security (Table 2.3; Fig. 2.5). Incorporation of NMs such as nanoclays, silver NPs, and ZnO into polymer matrices improves resistance to oxygen, carbon dioxide, and ethylene, reducing oxidation and microbial growth. These materials also provide UV protection, enhanced strength, and thermal stability, making them suitable for perishable foods like meat and dairy (Edo *et al.*, 2025). In beverage packaging, nanocomposites and nano-emulsions contribute to lighter yet stronger materials. Nanosensors and indicators enable real-time monitoring of food quality, such as detecting oxygen levels in vacuum-sealed products. Antimicrobial NMs, including CNTs, silver, and chitosan NPs, effectively inhibit pathogens (Ghosh *et al.*, 2025). Despite these benefits, concerns regarding toxicity, environmental impact, and human exposure remain. Therefore, strict regulatory evaluations, standardized testing, and clear risk communication are essential for the safe and sustainable application of nanotechnology in food packaging (Sahoo *et al.*, 2021).

## 2.5 Function of Nanoparticles in Sustainable Food Packaging

Conventional packaging materials are primarily polymers and typically have certain drawbacks when used solely for packaging fresh fruit flies (Sharma *et al.*, 2022). Over 350 million tons of plastic are manufactured worldwide yearly, with food packaging accounting for a significant portion of traditional plastic usage (Singh *et al.*, 2016; Sharma *et al.*, 2023). This puts a significant strain on the environment.

One approach that shows promise for addressing the drawbacks of conventional polymeric packaging made of plastic is the development of nanocomposites (Wu *et al.*, 2026). Owing to its numerous applications and potential uses, nanopackaging has been widely utilized for the sustainable preservation of fresh fruits and vegetables, as illustrated in Table 2.4 and Fig. 2.6. The application of

**Table 2.3.** Challenges of nanomaterials in sustainable food preservation and packaging.

| Challenges | Description | Examples / issues | References |
|---|---|---|---|
| Toxicity and health risks | Potential migration of nanoparticles into food can pose toxicity risks to human health due to accumulation in tissues and organs | Silver nanoparticles can accumulate and damage the liver or kidneys; the long-term exposure impact | Ferdous and Nemmar, 2020; Xuan *et al.*, 2023 |
| Environmental persistence | Many engineered nanomaterials are non-biodegradable and may accumulate in ecosystems, impacting soil and aquatic life | Nano-ZnO and $TiO_2$ can persist in soil and affect microbial activity | Kumar *et al.*, 2021b, 2022 |
| Regulatory uncertainty | Lack of standardized global regulatory frameworks hampers commercialization and safe use of nanomaterials in food systems | Different rules in the EU, USA, and Asia create confusion for developers and consumers | Onyeaka *et al.*, 2022 |
| Consumer acceptance | Lack of awareness or negative perceptions regarding nanotechnology in food can hinder public trust and market penetration | Many consumers are sceptical or reject nanolabeled food due to safety fears | Kuang *et al.*, 2020 |
| Cost of production and scaling | High cost of synthesis, purification and incorporation of nanomaterials increases production costs, limiting accessibility | Nano-encapsulation and active coatings are often costlier than traditional preservatives | del Rosario Herrera-Rivera *et al.*, 2024 |
| Stability during processing | Nanomaterials may lose functionality under food processing conditions such as high heat, pressure or pH changes | Lipid-based nanocarriers may degrade during pasteurization | |
| Detection and monitoring | Current analytical methods to detect and quantify nanomaterials in complex food matrices are limited and often costly or unreliable | Difficult to track nanoparticle migration or degradation inside multi-layered packaging | Villamayor *et al.*, 2023 |
| Standardization of materials | Lack of standard definitions and characterization methods complicates safety assessment and comparison across studies | No universal particle size threshold or classification system across nations | Sharifi *et al.*, 2022 |

*Continued*

**Table 2.3.** Continued

| Challenges | Description | Examples / issues | References |
|---|---|---|---|
| Functional compatibility | Nanomaterials may interact with food components, altering their intended function or triggering undesirable chemical reactions | Nano clay or carbon-based materials can absorb flavor or react with food acids | Chudasama and Goyary, 2024 |
| Waste disposal and recycling | Nanopackaging materials can interfere with current recycling streams or remain in landfill environments due to their inert nature | Carbon nanotubes and nanosilver residues may be unrecoverable and persist in waste | Živković *et al.*, 2014 |

nanotechnology in food technology is extensive and multifaceted. The NPs are commonly applied in two principal ways: they can be integrated directly into food items or disinfect food and packaging materials (Roychoudhury, 2020; Kumar, 2024). Within the food processing industry, NPs fulfill several roles, such as enhancing flavors, providing color, improving texture through gelling, preventing clumping as anticaking agents, serving as preservatives, and functioning as nanocarriers for the delivery of nutrients and other beneficial substances (Kumar, 2024; Wu *et al.*, 2026). Table 2.4 illustrates the role of nanotechnology in developing sustainable food packaging.

## 2.6 Nanotechnology for Safe, Sustainable, and Satisfactory Food Packaging

Nanotechnology presents significant advantages, but attention must also be directed toward the potential disadvantages of utilizing NMs in the packaging of fruits and vegetables (Biswas *et al.*, 2022). There is insufficient evidence to assert that handling materials developed through nanotechnology is either safer or more dangerous than their conventional counterparts in preserving postharvest produce (Ghosh *et al.*, 2025). Therefore, green nanotechnology is viewed as both a groundbreaking and eco-conscious approach, encouraging future studies to address and minimize the risks associated with products in the field of nanoscience (Gupta *et al.*, 2024). The three viewpoints regarding developing environmentally friendly, sustainable technologies for postharvest fruits and vegetables in this context. Implementing safe NMs calls for the standardization of NPs usage to promote sustainable practices in food preservation (Ahmad *et al.*, 2024). A research strategy is urgently needed to ensure proper regulation of NPs in this application, with three primary goals identified for achievement. This research aims to achieve three primary goals: (i) to formulate standardized guidelines for the characterization and quantification of NPs in postharvest fruits and vegetables through the use of standardized analytical

**Table 2.4.** Function and use of nanoparticles in sustainable food packaging and preservation.

| Nanoparticle type | Function in packaging/ preservation | Application example | Benefits | Reference |
|---|---|---|---|---|
| Silver nanoparticles (AgNPs) | Antimicrobial agent | Integrated into polymer films (e.g. PLA, PE) | Prevent microbial growth, prolong shelf life | Dube and Okuthe, 2025 |
| Zinc oxide nanoparticles (ZnO NPs) | UV-blocking, Antimicrobial, Gas barrier | Coated on biodegradable films for fruit packaging | Improve light stability, reduce microbial load, enhance film strength | Espitia *et al.*, 2012, 2025 |
| Titanium dioxide nanoparticles ($TiO_2$ NPs) | UV-blocking and self-cleaning | Used in packaging for dairy products | Protect against UV degradation, maintain nutrient integrity | Emamifar *et al.*, 2010 |
| Chitosan nanoparticles | Biopolymer nanocarrier for active compounds | Coating for strawberries and tomatoes | Biodegradable, antimicrobial, controls ripening | Yarahmadi *et al.*, 2024 |
| Silicon dioxide nanoparticles ($SiO_2$ NPs) | Moisture and gas barrier enhancement | Used in multi-layer packaging materials | Enhance packaging durability and transparency | Pulikkalparambil *et al.*, 2023 |
| Clay nanocomposites (e.g. montmorillonite) | Gas permeability reduction, mechanical strength | Applied in meat and cheese packaging | Improved shelf life through reduced oxygen exposure | Ninago *et al.*, 2024 |
| Starch-based nanocomposites | Biodegradable matrix, Barrier to gases and water vapor | Packaging for bakery items and dry food | Edible, renewable, and reduces plastic usage | Karnwal *et al.*, 2025 |
| Nano-encapsulated essential oils | Controlled release of natural antimicrobials and antioxidants | Citrus oil nano-emulsion in gelatin films | Enhances freshness, inhibits spoilage microbes | Gupta *et al.*, 2024 |
| Carbon nanotubes (CNTs) | Mechanical strength and electrical conductivity enhancement | Smart/intelligent packaging with temperature indicators | Signal spoilage, robust structure | Ramezani *et al.*, 2025 |
| Nanocellulose | Reinforcement and barrier improvement in biodegradable films | Edible coatings for fruits/ vegetables | Renewable, enhances strength and transparency | Azeredo *et al.*, 2017 |

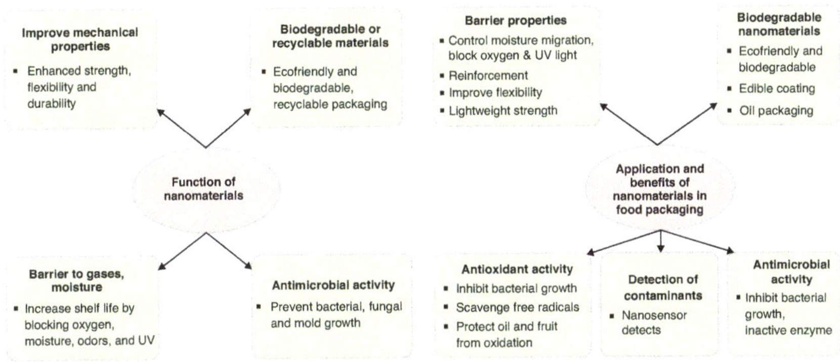


**Fig. 2.6.** Role of nanomaterials in sustainable food packaging and preservation.

techniques; (ii) to assess consumer exposure to NPs within the framework of postharvest food; and (iii) to establish a coherent toxicity assessment protocol for NPs on human health and environmental impact, in terms of sustainable nano-application, green nanotechnology advocates for the environmentally conscious use of NMs in the storage processes of postharvest food (Ghosh *et al.*, 2025). The production of green nanoproducts should ideally involve natural materials, such as plant secondary metabolites, which offer the advantages of being environmentally friendly, economically viable, biodegradable, and highly bioactive (Osman *et al.*, 2024; Espitia *et al.*, 2025).

Implementing nanoproducts in the preservation of fresh fruits and vegetables requires the recycling of leftover waste materials and the creation of new nanoproduct formulations. Incorporating specific plant extracts and essential oils should be advocated to augment the functional characteristics of nanoproducts, which can contribute to better quality preservation and prolonged shelf life of fresh fruits and vegetables throughout their storage period. Ensuring satisfactory quality is paramount, as highlighted by the aforementioned approaches (Alquraishi *et al.*, 2025). It is essential to prioritize safety, sustainable production practices, and the application procedures associated with nanoproducts. Various nanoproducts, including nanopackaging, nanocoating, and nano/microbubbles, can be utilized to preserve the quality and prolong the shelf life of postharvest fresh vegetables during storage (Alquraishi *et al.*, 2025; Ghosh *et al.*, 2025). A range of nanosensors can be employed to monitor quality and safety in real-time, providing alerts to production operators and markets, thereby enabling flexible adjustments in postharvest handling and shelf life to guarantee the availability of fresh foods (Operato *et al.*, 2025) (Fig. 2.7).

## 2.7 Conclusion

Nanotechnology presents transformative potential in addressing global food waste through precision, efficiency, and enhanced preservation across the food supply chain. By leveraging nanoscale materials and systems, significant

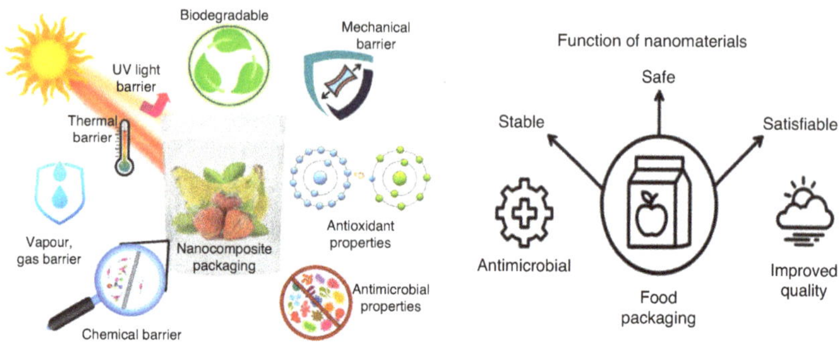


**Fig. 2.7.** Function of nanomaterials for safe, sustainable, and satisfactory food packaging.

advancements have been realized in extending the shelf life of perishable products, improving food packaging, and enabling real-time quality monitoring. Nano-enabled intelligent packaging systems, incorporating nanosensors and indicators, facilitate early detection of spoilage, microbial contamination, and physicochemical changes, thus reducing unnecessary disposal of still-edible food. Nanocoatings and edible films infused with antimicrobial and antioxidant agents minimize microbial proliferation and oxidative degradation, particularly in fruits, vegetables, and processed products. The application of nanotechnology in active packaging also contributes to controlled release of preservatives, moisture regulation, and barrier properties against oxygen, ethylene, and light factors that critically influence product longevity. In food processing, nano-emulsions and nanocarriers enhance the stability and bioavailability of functional compounds, reducing nutrient loss and improving the value of food products, thereby mitigating waste originating from nutrient degradation and spoilage. The role of nanotechnology in logistics and distribution, via smart labels and radio frequency identification (RFID) systems integrated with nanosensors, fosters a more responsive and efficient supply chain that can dynamically adjust to food conditions and reduce inventory losses.

However, while nanotechnology offers promising solutions, sustainability also demands thorough evaluation of NMs' environmental and health implications throughout their life cycle. Comprehensive risk assessments and development of regulatory frameworks are essential to ensure safe integration into food systems. In summary, combining nanotechnology and sustainability strategies provides a multifaceted approach to food waste reduction, aligning with global efforts toward circular economy principles and food security enhancement. Continued interdisciplinary research and responsible innovation are imperative to harness the full potential of nanotechnology while ensuring environmental safety and socio-economic equity.

# 3

# Advanced Smart Sensors and Nanobiosensors for Enhanced Quality Monitoring

## 3.1 Introduction

Smart sensors and nanobiosensors are pivotal in advancing quality control and safety in food nanotechnology by enabling precise, real-time monitoring of complex food systems. These devices integrate NMs with sensing components to detect physical, chemical, and biological parameters with high sensitivity and specificity (Sharma *et al.*, 2023b; Yadav *et al.*, 2025). The incorporation of nanostructured ingredients and nanoscale packaging has increased system complexity, necessitating intelligent diagnostic tools for maintaining product integrity and traceability across the supply chain (Sharma *et al.*, 2023). Smart sensors function by converting environmental stimuli into measurable signals and can be integrated with microcontrollers and wireless networks for remote monitoring. In food applications, they detect key parameters such as temperature, humidity, pH, gas emissions (e.g. ethylene and ammonia), and microbial activity. When embedded in packaging, they form intelligent systems capable of indicating freshness, identifying spoilage, and supporting cold chain management through time–temperature integrators (TTIs; Sharma *et al.*, 2023; Awlqadr *et al.*, 2024; Qadir, 2025).

Nanobiosensors further enhance detection capabilities by combining biological recognition elements such as enzymes, antibodies, nucleic acids, and aptamers with nanostructured transducers (Ahmed *et al.*, 2022; Sharma and Dhadly, 2023). These systems specifically identify target analytes, including pathogens, toxins, allergens, and contaminants. Advanced NMs such as gold nanoparticles (AuNPs), CNTs, graphene, quantum dots, and metal–organic frameworks (MOFs) enhance sensor performance through their high surface area and unique electronic properties (Kumar, 2024; Pengsomjit *et al.*, 2024). Consequently, nanobiosensors provide ultralow detection limits, rapid response times, and multiplex detection. In food processing, these technologies enable continuous monitoring of critical control points, ensuring that NM-based additives and preservatives remain within safe regulatory limits, thereby enhancing food safety and quality assurance (Ghosh *et al.*, 2022).

Nanobiosensors are advanced analytical tools widely used in food quality control and safety monitoring. They combine biological recognition elements such as enzymes, antibodies, nucleic acids, and aptamers with nanostructured transducers to achieve high sensitivity, specificity, and low detection limits (Ramesh *et al.*, 2022; Nath, 2024). These bioreceptors selectively bind to target analytes, including foodborne pathogens (e.g. *Salmonella*, *Listeria*), toxins, allergens, and environmental contaminants like pesticides and heavy metals (Rani *et al.*, 2025b; Yang *et al.*, 2025). Nanomaterials such as AuNPs, CNTs, graphene, quantum dots, and MOFs play a crucial role in enhancing biosensor performance. Their high surface area allows efficient immobilization of biomolecules, while unique optical and electrical properties enable signal amplification and rapid response (Kwon and Ko, 2022). For instance, graphene supports fast electron transfer in electrochemical sensing, AuNPs enhance optical detection via surface plasmon resonance (SPR), and quantum dots enable multiplexed detection through tunable fluorescence.

These features make nanobiosensors highly effective for real-time, on-site monitoring across the food supply chain. They can detect contaminants at extremely low concentrations (ppb or ppt levels), enabling early intervention and prevention of foodborne outbreaks (Kumari *et al.*, 2022). Furthermore, integration into smart sensor systems allows continuous monitoring of critical control points during food processing. Such systems can track parameters like temperature, pH, humidity, and chemical composition, ensuring optimal processing conditions and regulatory compliance (Fig. 3.1). Nanobiosensors provide rapid, accurate, and cost-effective solutions for improving food safety and quality (Chong *et al.*, 2023). Nanobiosensors play a vital role in monitoring advanced food systems (Table 3.1). For example, nanocapsules delivering antioxidants or antimicrobials must be tracked to ensure stability and avoid harmful degradation. Smart pH sensors in fermentation processes help optimize conditions, improving yield and consistency. These sensors are

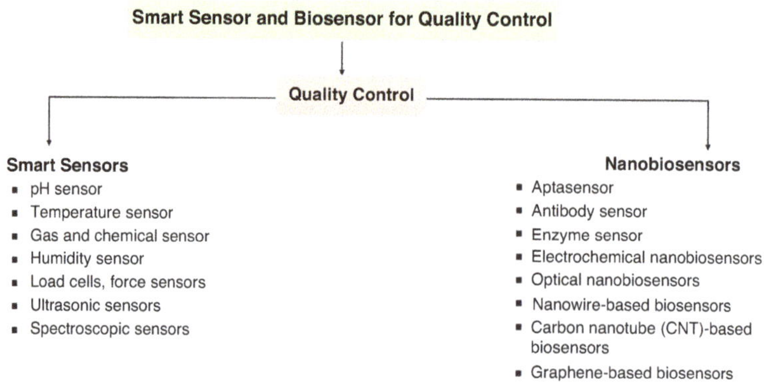


**Fig. 3.1.** Smart sensors and nanobiosensors for quality control.

**Table 3.1.** Classification and applications of intelligent sensors and nanobiosensors in food nanotechnology.

| Sensor type | Nanomaterials/components | Mechanism of detection | Application in the food sector | Advantages | References |
|---|---|---|---|---|---|
| Optical nanobiosensors | Quantum dots, AuNPs, Graphene oxide | Fluorescence resonance energy transfer (FRET), colorimetric | Detection of pathogens (*E. coli, Salmonella*), *mycotoxins* | High sensitivity, real-time detection | Rani *et al.*, 2025; Yang *et al.*, 2025 |
| Electrochemical biosensors | Carbon nanotubes (CNTs), ZnO NPs, AuNPs | Current/voltage change due to bio-recognition | Pesticide residues, heavy metal ions, and glucose monitoring | High specificity, portable | González Alejandre *et al.*, 2021; Suresh *et al.*, 2023 |
| Piezoelectric biosensors | ZnO nanowires, Quartz crystal | Mass change detection via resonance frequency shift | Bacterial contamination, spoilage detection | Label-free, real time | Awlqadr *et al.*, 2024 |
| Thermal biosensors | Metal-oxide nanoparticles, Polymer-based nanomaterials | Heat change upon analyte binding | Enzyme activity, freshness indicators | High stability | Ramesh *et al.*, 2022 |
| Colorimetric sensors | AuNPs, AgNPs, $MoS_2$ nanosheets | Color change via aggregation or oxidation | Smart packaging, freshness indicators (meat, fish) | Visual and easy-to-use | Kwon and Ko, 2022 |
| Gas/vapor sensors | Graphene, Metal-organic frameworks (MOFs), $SnO_2$ NPs | Resistance changes upon gas adsorption | Detection of spoilage gases ($H_2S$, $NH_3$, ethanol) | Early spoilage detection | Wang *et al.*, 2016 |
| RFID-integrated sensors | Flexible nanomaterials, Antenna with AuNPs | Wireless signal variation | Temperature and humidity monitoring in food logistics | Noninvasive, remote tracking | Zuo *et al.*, 2022 |
| Aptasensors | DNA/RNA aptamers with AuNPs, graphene oxide | Binding-induced signal generation | Allergen and toxin detection | High specificity and stability | Wang *et al.*, 2025 |
| Enzyme-based nanobiosensors | Enzyme-functionalized CNTs, $TiO_2$, chitosan | Substrate interaction causes signal change | Sugar and alcohol content analysis | Highly selective | Qu *et al.*, 2025 |
| Immunosensors | Antibody-modified AuNPs, magnetic NPs | Antigen-antibody interaction | Pathogen detection (*Listeria, Salmonella*) | High affinity and selectivity | Janik-Karpinska *et al.*, 2022 |

highly effective in detecting foodborne pathogens such as *Salmonella*, *Listeria monocytogenes*, and *Escherichia coli*, enabling rapid, on-site analysis without complex preparation (Rani *et al.*, 2025; Yang *et al.*, 2025). Electrochemical nanobiosensors enable rapid, real-time, and label-free detection of bacterial DNA or associated toxins, whereas optical sensing approaches such as SPR and fluorescence resonance energy transfer (FRET) offer high-speed, multiplexed analytical capabilities (Zuo *et al.*, 2022; Wang *et al.*, 2025).

## 3.2 Principles of Nanobiosensors

Nanobiosensors are advanced analytical devices that integrate nanotechnology and biotechnology to detect specific biological, chemical, or physical analytes with high precision. Their operation relies on three fundamental components: a biorecognition element, a transducer, and a signal processing system (Song *et al.*, 2025). The biorecognition element, such as enzymes, antibodies, nucleic acids, or synthetic receptors, selectively interacts with the target analyte, ensuring high specificity and minimal interference from other substances (Kulkarni *et al.*, 2022). This selectivity is critical for accurate detection in complex matrices. The transducer converts the bio-recognition event into a measurable signal via physicochemical mechanisms, such as electrochemical, optical, piezoelectric, or thermal changes. This conversion enables the transformation of molecular interactions into quantifiable outputs suitable for analysis. To enhance sensitivity, nanobiosensors often incorporate signal amplification strategies, such as NM-assisted enhancement, enzymatic catalysis, or advanced electronic and optical amplification techniques. These approaches significantly improve the signal-to-noise ratio, allowing detection of trace-level analytes (Rahman *et al.*, 2022; Kakimova *et al.*, 2023).

The generated signals are processed through integrated readout systems, including electronic devices, optical detectors, or imaging platforms, which translate raw signals into interpretable data (Kim *et al.*, 2024). A key advantage of nanobiosensors is their miniaturization and integration capability. The use of nanoscale materials enables the development of compact, portable, and highly efficient devices that can be incorporated into lab-on-chip systems, wearable technologies, and smart packaging (Kulkarni *et al.*, 2022). The nanobiosensors offer rapid, sensitive, and real-time monitoring capabilities, making them highly valuable in environmental monitoring, healthcare diagnostics, food safety, and industrial applications, where accuracy, speed, and adaptability are essential (Kumar *et al.*, 2024; Song *et al.*, 2025).

## 3.3 Working Mechanisms of Nanobiosensors

Nanobiosensors work based on various techniques in which NMs and biological recognition elements are used to detect and analyze target analytes. The working mechanism of a nanobiosensor depends on its design and configuration (Chen *et al.*, 2023b, 2024). Some typical applications are optical detection, electrochemical detection, piezoelectric detection, magnetic detection, and nanomechanical detection. Optical, electrochemical, piezoelectric, magnetic, and nanomechanical detection techniques (Sharma *et al.*, 2023). Changes in the optical properties of these NMs, such as fluorescence intensity or SPR, occur when the detection element binds to the analyte (Bhardwaj *et al.*, 2022). On the other hand, electrochemical nanobiosensors lead to changes in electrical properties, such as conductance or impedance changes, used to investigate binding events (Chen *et al.*, 2022b). Piezoelectric nanobiosensors use NMs with piezoelectric properties, causing changes or deformations in mechanical stress due to binding phenomena; magnetic nanobiosensors use magnetic NMs, where changes in magnetization or magnetic field response indicate that the analyte binds to nanobiosensors they are measuring structures like mass or resonance frequency to detect binding events, to determine bonding phenomena (Chen *et al.*, 2022a; Song *et al.*, 2025). These working mechanisms show the various approaches nanobiosensors use to detect and measure target analytes (Chen *et al.*, 2022). The distinctive properties of NMs and their interaction with recognition elements, nanobiosensors offer high sensitivity and selectivity in detection. They can be customized for specific analytes and applications, enabling real-time monitoring of essential parameters in food packaging and shelf-life management (Ghosh *et al.*, 2022)

## 3.4 Nanobiosensors for Food Packaging

Nanobiosensors are advanced analytical devices that integrate nanotechnology and biotechnology for real-time monitoring of food quality, safety, and shelf life. These sensors can be embedded within packaging systems or external devices to continuously track critical parameters such as temperature, humidity, gas composition, and microbial activity (Valenzuela-Amaro *et al.*, 2023). By providing precise and dynamic data, nanobiosensors enable timely interventions and controlled processing conditions, thereby maintaining food freshness and safety. They operate through optical, electrochemical, or nanomechanical detection mechanisms to identify specific analytes, including spoilage indicators and pathogenic microorganisms (Chen *et al.*, 2023). Their integration into intelligent packaging systems enhances food integrity by detecting deviations such as temperature fluctuations, poor ventilation, or microbial contamination, allowing corrective measures to be implemented promptly (Kumar *et al.*, 2021a; Abdel-Karim, 2024).

The advantage of nanobiosensors is their ability to ensure early detection of contamination risks, thereby preventing the distribution and consumption of unsafe food products. They play a crucial role in accurate shelf-life prediction by continuously monitoring environmental and biochemical changes within packaged food (Fig. 3.2; Abdel-Karim, 2024). This facilitates improved inventory management and reduces unnecessary

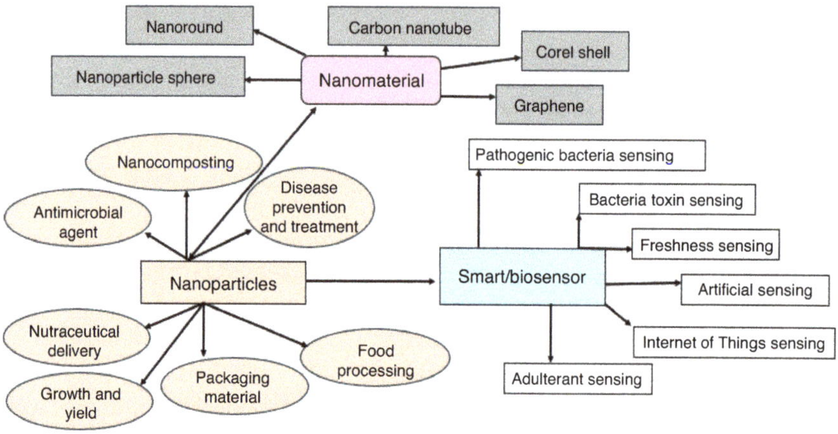


**Fig. 3.2.** Applications of nanoparticles in the food industry with a focus on nano/biosensors for ensuring food authenticity and safety, and the types of nanomaterials used.

food waste. Furthermore, nanobiosensors enhance traceability and transparency in the food supply chain through integration with radio-frequency identification (RFID) or QR-based systems, enabling access to product history and storage conditions. Their application supports sustainability by minimizing food loss and optimizing resource utilization (Halonen *et al.*, 2020).

## 3.5 Types of Nanosensors

Nanosensors offer promising new ways to monitor the quality and safety of food in the agricultural and food industries (Jafarizadeh-Malmiri *et al.*, 2019). A practical and cutting-edge piece of analytical equipment, a biosensor utilizes physical, chemical, and biological sensing techniques. It can be deduced that nanobiosensors are used in the detection of urea, glucose, pathogens, and varieties of microorganisms because the objective of biosensors is to find out the diversity existing in biomolecular complexes such as oligonucleotides, antigen–antibody, living cells, tissue-level interactions, and interactions with both enzyme substrate and lectin–glycoprotein (Aggarwal *et al.*, 2021; Chen *et al.*, 2023). Nanosensors are favored over other sensors because of their convenient size, durability, mobility, reproducibility, and stability. An inert, biocompatible matrix component monitors with or without an encapsulated optical reporter (Jafarizadeh-Malmari *et al.*, 2019). When it comes to nanobiosensors, the NMs are the backbone of the transducer because so many variables are reliant on the limited physical nanoscale, nanofabrication of nanoparticles, or nanostructured surfaces (Srivastava *et al.*, 2018; Kumar *et al.*, 2024).

Sensors (transducers) are made out of NMs like quantum dots, graphene oxide, carbon nanotubes (CNs), and metallic AuNPs (Srivastava *et al.*, 2018; Singh, 2020; Singh *et al.*, 2023a). Nanoconfinement, morphological flexibility, improved optical, mechanical, electrical, and thermal properties, and a high specific surface area contribute to a NM's improved transducing capabilities (Srivastava *et al.*, 2018). This classification of nanosensors and nanobiosensors is based on their output-generating mechanism, which is based on the transduction mechanism (Srivastava *et al.*, 2018). Nanomaterials or CNT-based electrodes are connected to electrochemical sensors. Electrode configurations based on CNT are linked to electrochemical sensors. The metallic nanoparticle can connect to other nanoparticles, such as upconversion NMs and quantum dots in optical nanosensors, to improve their efficiency and versatility, so that it can take advantage of its increased optical feature (Srivastava *et al.*, 2018; Monavari *et al.*, 2023). The mechanical and piezoelectric capabilities of microcantilevers and crystals place some mass nanosensors in a third group with limited applications in the food and agricultural industries (Srivastava *et al.*, 2018; Monavari *et al.*, 2023).

### 3.5.1 Electrochemical sensors

The chemical and biomolecular nanofabricants react with the biological components and the analytes of interest. Electrochemical signals (voltage, current, and impedance) can quantify the amount of electron gain or loss during biointeractions (Srivastava *et al.*, 2018). These electrochemical signals are generated quantitatively and related to an analyte's presence in a sample. As reported by Srivastava *et al.* (2018), compatible with present miniaturization and nanofabrication technologies, minimal energy requirements, reliability, relatively inexpensiveness, minimal maintenance, speed, low detection limits, and simplicity are essential for the practical application of electrochemical transducers' extreme sensitivity (Srivastava *et al.*, 2018).

According to Srivastava *et al.* (2018), electrochemical nanosensor devices can be divided into three groups based on their underlying measurement principle: amperometry, voltammetry, and potentiometry. An amperometric sensor is a form of electrochemical sensor that continuously measures the current produced by redox reactions of electroactive species in a biochemical reaction (Srivastava *et al.*, 2018). At a specific voltage, the Faradaic current determines the concentration of electroactive species. Current is typically linear with respect to voltage, forming a proportion with electroactive species (Srivastava *et al.*, 2018). Amperometric nanosensors, such as the adenosine triphosphate (ATP) development sensor (Kuang *et al.*, 2020) and cutting-edge pregnancy tests based on the beta-HCG (human chorionic gonadotropin) sensors, are widely used in the healthcare and diagnosis industries (Malhotra and Chaubey, 2003).

Electrochemical nanosensors are widely applied in the food industry because of their high
sensitivity, rapid response, and ability to detect contaminants, pathogens, toxins, and quality markers even at trace levels. Below are some examples of nanosensors (specifically eletrochemical types) currently used in the food sector:

### A. Graphene-based Electrochemical Nanosensors

Detection of foodborne pathogens such as Salmonella and E. coli. For example, Graphene oxide-modified electrodes are used to detect microbial contamination in milk and meat products.

### B. Gold Nanoparticle (AuNP)-based Electrochemical Nanosensors

Monitoring pesticide residues (e.g., organophosphates) in fruits and vegetables. For example, AuNP-modified electrodes for detecting chlorpyrifos and malathion in fresh produce.

### C. CarbonNanotube (CNT)-based Nanosensors

Determination of mycotoxins (e.g., aflatoxin B1) in cereals, nuts, and dairy. For example, CNT-modified glassy carbon electrodes for rapid aflatoxin detection in peanuts.

### D. MetalOxide Nanoparticle-based Sensors (ZnO, TiO2, Fe3O4)

Monitoring food freshness through detection of biogenic amines (putrescine, histamine, cadaverine).Forexample,ZnOnanoparticle-basedelectrochemicalse nsorsforfishandmeat spoilage detection.

### E. Nanocomposite-based Electrochemical Biosensors

Detectionofallergenssuchasgluten,peanutproteins,andmilkproteins. Forexample, AuNP/graphene nanocomposite electrodes for allergen detection in bakery products.

### F. QuantumDot (QD)-based Electrochemical Nanosensors

Detectionofheavymetals(Pb2+,Cd2+,Hg2+)inbeveragesandwaterusedinfood-processing.For example, CdS quantum dot-modified sensors for lead detection in soft drinks.

### G. Molecularly Imprinted Polymer (MIP)-based Nanosensors (with nanomaterials)

Detectionofantibiotics(tetracycline,penicillin,sulfonamides)inmilkand-meat.Forexample, MIP-AUNP sensors for tetracycline monitoring in dairy products.

### 3.5.2 Optical sensors

Optically based biosensors are very compatible with a wide range of spectroscopic measurements, including absorption, phosphorescence, fluorescence, Raman, and surface-enhanced Raman scattering (SERS) (Antiochia *et al.*, 2013; Srivastava *et al.*, 2018), and are used to detect changes in the optical signal. The metrics above are determined by sensing the variations in wavelength, phase, intensity, time, and polarity of light (Srivastava *et al.*, 2018; Song *et al.*, 2025). The notion of quantum dots based on sensing systems has advanced with selective functionalization, paving the way for their application in agro-farming and the food processing industry (Srivastava *et al.*, 2018; Ameta *et al.*, 2020). This appeared to be an interesting strategy. For example, the silane group is utilized to modify the surfaces of CdSe and ZnS quantum dots (Srivastava *et al.*, 2018). Moreover, for the detection of dicyandiamide in the milk product, these CdSe and ZnS quantum dots are also coupled with a methylacrylate-functionalized molecularly imprinted polymer to create an optosensor based on quantum dots (Liu *et al.*, 2021). The quantum dots detect fluorescence quenching, which is linearly related to the concentration of dicyandiamide (Liu *et al.*, 2021).

## 3.6 Nanosensor Application in Innovative Food Packaging

Food packaging plays a vital role in maintaining product quality and ensuring safety during storage and distribution in the food processing sector. It directly influences shelf life, organoleptic characteristics, and nutritional value (Aggarwal *et al.*, 2021; Chen *et al.*, 2022). However, conventional packaging systems exhibit limitations in effectively controlling spoilage and monitoring dynamic environmental factors such as temperature, pH, moisture, humidity, and gas composition. Recent advancements have led to the development of active and bioactive packaging systems capable of regulating these parameters, thereby enhancing product stability and safety (Chen *et al.*, 2022; Kuppusamy *et al.*, 2024). Despite these innovations, increasing concerns regarding food contamination and bioterrorism necessitate the adoption of rapid, sensitive, and reliable monitoring technologies. Traditional analytical methods, although widely used, often lack real-time detection capability and are becoming insufficient for modern food safety requirements. Biosensors have emerged as promising tools for addressing production and storage challenges. These analytical devices integrate biological recognition elements such as enzymes, antibodies, nucleic acids, aptamers, and microorganisms with transducers to detect and quantify specific analytes (Aggarwal *et al.*, 2021; Guruprasath *et al.*, 2024).

Various types of biosensors have been developed for food applications, including electrochemical biosensors for detecting redox reactions, optical biosensors for identifying pesticide residues and mass-sensitive biosensors for monitoring chemical changes (Aggarwal *et al.*, 2021; Guruprasath *et al.*,

2024). However, despite their high sensitivity and specificity, biosensors often require relatively large sample volumes, which limits their integration into intelligent packaging systems (Biswas *et al.*, 2022). Conventional chemical and enzymatic sensors, while accurate, are often expensive and operationally complex. These limitations highlight the need for more cost-effective, miniaturized, and real-time sensing technologies for advanced food safety monitoring (Kumar *et al.*, 2024).

Hu *et al.* (2024) developed a label-free electrochemical immunosensor based on a functionalized nanofiber membrane incorporating polyvinyl alcohol/gelatin-stabilized gold nanoparticles (PVA/G-AuNPs) for the detection of quinalphos, a colorless organophosphate insecticide. The sensor exhibited high sensitivity, with a wide linear detection range of 0.06–1000 ng/ml and a low detection limit of 50.74 pg/ml ($S/N = 3$), demonstrating its applicability for food safety monitoring. Similarly, Atilgan *et al.* (2023) fabricated an electrospun nanofiber-based enzymatic biosensor for monosodium glutamate (MSG) detection using PCL–CHIT/PAMAM G1–Mt/GluOx nanocomposites. This system integrates polycaprolactone, chitosan, dendrimer-modified montmorillonite, and glutamate oxidase to achieve high specificity and sensitivity. The biosensor enabled accurate MSG quantification in tomato soup without matrix interference, showing a linear range of 0.0025–0.175 mM, a detection limit of 1.045 μM and a recovery efficiency of 103.125%.

Shavisi (2024) developed electrospun chitosan–carrageenan (CH–KC) nanofiber mats incorporating anthocyanins extracted from *Malva sylvestris* for intelligent food packaging. These bioactive films exhibited desirable mechanical and barrier properties, including tensile strength (7.71–11.02 MPa) and controlled permeability. Anthocyanin-based systems function as pH-sensitive indicators, enabling real-time freshness monitoring, as demonstrated in carp fillets. Tavana *et al.* (2024) introduced kafirin-polyethylene oxide (PEO) bio-nanocomposite pH indicators using beetroot extract, while Hassan *et al.* (2024) designed a bacterochromic cellulose biosensor with immobilized anthocyanins on electrospun cellulose nanofibers. This system detects microbial activity via a hypsochromic shift from 595 nm (purple) to 448 nm (green), indicating lactic acid production by *Lactobacillus acidophilus*. Collectively, these innovations highlight the potential of nanofiber-based biosensors and intelligent packaging systems for advanced food quality monitoring. Advances in nanotechnology have enabled the integration of nanosensors into food packaging systems, significantly enhancing their functionality and reliability. Nanosensors facilitate real-time monitoring of food quality across various stages of processing, storage, and distribution, thereby ensuring product safety and integrity (Monavari *et al.*, 2023). They help prevent exposure to external factors such as oxygen, moisture, dust, radiation, and microorganisms, which can deteriorate the organoleptic and nutritional properties of both dried and nondried food products (Monavari *et al.*, 2023).

In smart packaging systems, nanosensors play a crucial role in maintaining freshness, detecting tampering, and providing information on production and expiration dates, as well as ensuring biocompatibility of packaging materials. These systems also support traceability and transparency, improving consumer confidence. Compared to conventional methods such as RFID, nanosensors offer enhanced sensitivity and the ability to monitor physicochemical and biological changes at the nanoscale (Chen *et al.*, 2022; Alshammari, 2023). The application of nanosensors in innovative and intelligent packaging has gained widespread attention in the food industry due to their potential to extend shelf life, maintain quality, and ensure accurate product labeling. Commercial nanosensor-based systems are already available for monitoring parameters such as time–temperature history and freshness indicators. Nanosensors enable the detection of allergens, pathogenic microorganisms, and chemical contaminants, thereby improving food safety (Chen *et al.*, 2022; Alshammari, 2023). Nanosensors can assess both external environmental conditions and internal changes within packaged foods, including oxygen levels, moisture content, and microbial activity. Integration of nanobiosensors into packaging materials allows early detection of leakage, spoilage, and microbial growth. Nanosensor-based packaging represents a transformative approach for enhancing food safety, quality control, and supply chain management (Gupta *et al.*, 2024b).

## 3.7 Smart Sensors in Food Quality Control

Smart sensors are crucial in improving food quality control by enabling real-time monitoring, automated inspection, and early detection of potential issues. They can track temperature, humidity, gas levels, and microbial growth, helping maintain optimal food storage, processing, and transportation conditions. Here's how smart sensors contribute to food quality control (Tutul *et al.*, 2023).

### 3.7.1 Real-time monitoring

Real-time smart sensor monitoring has emerged as a transformative strategy for ensuring food quality and safety across the supply chain. These advanced systems enable rapid, on-site, and nondestructive assessment of critical physical, chemical, and biological parameters associated with food freshness, contamination, spoilage, and nutritional integrity (Chen *et al.*, 2022). Integrated with wireless communication, Internet of Things (IoT) platforms, and cloud-based analytics, smart sensors provide continuous, highly sensitive monitoring tailored to modern processing and storage environments. Smart sensor technologies encompass gas sensors, biosensors, electrochemical sensors, optical sensors, and temperature/humidity sensors. Gas sensors detect volatile organic compounds (VOCs) released during spoilage of meat and fish, while biosensors identify foodborne pathogens such as *Salmonella*, *E. coli*, and *L. monocytogenes* through specific biological recognition elements (Fuertes *et al.*, 2016b; Kuppusamy *et al.*, 2024).

These approaches offer rapid and accurate detection, significantly reducing reliance on time-consuming conventional laboratory methods. The incorporation of NMs, including CNTs, graphene, and metal oxide nanoparticles, has further enhanced sensor sensitivity, selectivity, and miniaturization. Nano-enabled sensors can detect trace levels of contaminants such as pesticides, allergens, and mycotoxins (Dhal and Kar, 2025; Rani *et al.*, 2025). Optical sensors based on fluorescence and SPR enable label-free detection, whereas electrochemical sensors convert chemical interactions into measurable electrical signals for real-time analysis. In food logistics, particularly cold chain systems, smart sensors ensure strict control of temperature and humidity, with instant alerts for deviations. Smart labels such as TTIs and RFID tags improve traceability, support regulatory compliance, and contribute to waste reduction by enabling accurate shelf-life prediction (Table 3.2; Dhal and Kar, 2025).

### 3.7.2 Rapid detection

Rapid detection smart sensors have emerged as a transformative approach in food quality control, enabling real-time monitoring, rapid decision making, and enhanced food safety across the supply chain (Dong *et al.*, 2024; Daramola *et al.*, 2025). These sensors employ diverse detection mechanisms, including electrochemical, optical, piezoelectric, and biosensing technologies, to identify contaminants, spoilage indicators, and nutritional attributes. Unlike conventional methods, which are labor-intensive, time-consuming, and require sophisticated instrumentation, smart sensors provide rapid, cost-effective, portable, and user-friendly solutions suitable for on-site applications and automation (Dong *et al.*, 2024; Daramola *et al.*, 2025). Smart sensors operate by responding to specific physical, chemical, or biological stimuli and converting them into quantifiable signals. Electrochemical sensors are widely used for detecting pH, ions, and toxicants such as heavy metals and pesticides. Optical sensors, including fluorescence and SPR-based systems, enable label-free and real-time detection of microbial contamination and chemical residues (Dong *et al.*, 2024). Biosensors, incorporating biological recognition elements such as enzymes, antibodies, or nucleic acids, offer high specificity and sensitivity for detecting foodborne pathogens like *Salmonella*, *E. coli*, and *L. monocytogenes*, as well as allergens. The integration of NMs, such as CNTs, quantum dots and metal nanoparticles, further enhances sensor sensitivity, stability, and signal transduction efficiency (Inês and Cosme, 2025; Ma *et al.*, 2023). The integration of IoT and wireless communication technologies has enabled the development of intelligent sensor networks for continuous monitoring during storage, transportation, and retail. These systems track key parameters such as temperature, humidity, gas emissions (e.g. ethylene and ammonia), and VOCs. Additionally, smartphone-based platforms combined with AI and machine learning facilitate real-time data analysis, enabling predictive assessment of spoilage, improved shelf-life estimation, and enhanced traceability in modern food systems (Sharma and Shivandu, 2024).

**Table 3.2.** Major types of smart sensors, their mechanisms, target analytes, and food applications.

| Type of smart sensor | Mechanism | Target analytes | Food applications | References |
|---|---|---|---|---|
| Gas sensors | Chemiresistive, electrochemical, optical sensing | Ethylene, ammonia, hydrogen sulfide, $CO_2$, VOCs | Monitoring ripeness (fruits, vegetables), spoilage (meat, fish) and packaging atmosphere | Ma *et al.*, 2023; Rani *et al.*, 2025 |
| Biosensors | Biorecognition (enzyme/antibody) + transducer (electrochemical/optical) | Pathogens (*E. coli*, *Salmonella*), toxins, allergens, pesticides | Microbial safety monitoring, detection of allergens, antibiotic residues | Inês and Cosme, 2025 |
| Nanobiosensors | Nanostructured materials (AuNPs, CNTs, QDs) + biological recognition | Heavy metals, mycotoxins, foodborne pathogens, drug residues | Enhanced sensitivity for trace contaminants, on-site food testing | Singh *et al.*, 2025a |
| Colorimetric sensors | pH-sensitive or redox-responsive dyes | pH changes, spoilage metabolites | Visual spoilage indicators in intelligent packaging | Kwon and Ko, 2022 |
| Electronic nose (E-nose) | Array of chemical sensors + pattern recognition software | VOCs, aroma compounds | Quality control, freshness assessment, adulteration detection | Wilson, 2018; Chen *et al.*, 2022; Vanaraj *et al.*, 2025 |
| Electronic tongue (E-tongue) | Array of taste sensors + signal processing | Taste compounds (sugar, salt, bitterness) | Taste profiling, quality grading of beverages, dairy, oils | Vanaraj *et al.*, 2025 |
| RFID-based smart sensors | RFID + sensing element (temperature, gas, humidity) | Temperature, humidity, gas composition | Cold chain monitoring, anticounterfeiting, traceability | Zuo *et al.*, 2022 |
| Optical sensors (fluorescence, SPR) | Light-based detection (change in refractive index or emission) | Toxins, pathogens, and environmental parameters | Real-time detection of foodborne hazards, freshness | Boro *et al.*, 2025; Daramola *et al.*, 2025 |
| Capacitive/micro-electro-mechanical systems (MEMS) sensors | Change in capacitance due to analyte interaction | Moisture, strain, gas, biological molecules | Food packaging, structural health and freshness | Jayan *et al.*, 2025 |
| Infrared (IR) sensors | Absorption of IR radiation by molecules | Water content, fat, sugar | Nutritional profiling, quality control, food grading | Fodor *et al.*, 2024 |
| Magnetic biosensors | Magnetic nanoparticles + target interaction | Bacteria, toxins, DNA | Ultra-sensitive pathogen detection | Dong *et al.*, 2024 |

### 3.7.3 Ultrasensitive detection

Ultrasensitive detection in food quality control has become increasingly important due to the growing demand for food safety, authenticity and regulatory compliance. This analytical approach allows identifying and quantifying trace-level contaminants, pathogens, allergens, adulterants, and chemical residues that could pose significant health risks to consumers (Dong *et al.*, 2024). The evolution of detection technologies, such as biosensors, immunoassays, chromatography, and spectrometry, has contributed significantly to advancing ultrasensitive analytical techniques (Chen *et al.*, 2023). Moreover, integrating nanotechnology, microfluidics and lab-on-a-chip systems has revolutionized detection capabilities by enabling rapid, precise, on-site analyses. The application of ultrasensitive detection methods is critical in identifying harmful agents such as pesticides, mycotoxins, heavy metals, veterinary drugs, and microbial toxins in food products (Boshra *et al.*, 2024). Traditional methods such as gas chromatography (GC) and liquid chromatography (LC) coupled with mass spectrometry (MS) are widely used for their high sensitivity and specificity. However, these techniques often require complex sample preparation and are time-consuming. Recent advancements, including SERS, enzyme-linked immunosorbent assay (ELISA), and aptamer-based biosensors, offer rapid detection with high sensitivity and low detection limits (Boshra *et al.*, 2024).

### 3.7.4 Specific target detection

Specific target detection in food quality control is vital to ensuring food safety, authenticity, and nutritional value. It involves identifying and quantifying specific analytes, including pathogens, allergens, toxins, pesticides, genetically modified organisms (GMOs), and spoilage indicators (Nath, 2024). These targets may be biological, chemical, or physical, and their detection often requires sensitive and selective analytical tools. As the global food industry faces increasing pressure to deliver safe and high-quality products, developing advanced detection technologies has become paramount (Yosri *et al.*, 2024). Recent advances in nanotechnology and biotechnology have further enhanced the capabilities of food quality control systems (Wójcik *et al.*, 2025). Nanomaterials such as AuNPs, quantum dots and CNTs are increasingly integrated into biosensors to improve detection limits (Camargo *et al.*, 2024). These materials offer high surface area, excellent conductivity and tunable optical properties, making them ideal for signal amplification. Aptamer-based sensors are emerging as practical tools for detecting specific targets due to their high binding affinity and stability (Zhang *et al.*, 2009; Konfo *et al.*, 2024).

### 3.7.5 Rapid response

Rapid response in food quality control is essential to ensure food safety, minimize health risks, and maintain consumer confidence. With globalization, rising food demand, and frequent outbreaks of contamination, the food industry requires rapid, reliable, and sensitive analytical tools (Visciano and Schirone, 2020; Duan *et al.*, 2025). Traditional methods such as culture-based microbiological techniques, chromatography, and spectroscopy are reliable but often time-consuming and labor-intensive. Hence, the focus has shifted to developing and deploying rapid response technologies that can detect contaminants, pathogens, and spoilage in real-time or near real-time (Chen *et al.*, 2022). Technological advancements have enabled biosensors, nanotechnology, and digital innovations for prompt food quality assessment. Biosensors offer specificity and sensitivity, particularly those based on immunological and enzymatic reactions (Leslie *et al.*, 2025). With minimal sample preparation, these devices can detect microbial contamination, pesticide residues, allergens, and adulterants (Xu *et al.*, 2025).

### 3.7.6 Enhanced shelf life

Enhanced shelf-life smart sensors are revolutionizing food quality control by real-time monitoring of freshness, safety and environmental conditions during storage and transportation. These sensors integrate nanotechnology, materials science and electronics advances to ensure that food remains safe and consumable over extended periods. Their application is crucial in reducing food spoilage, improving traceability, and minimizing food waste (Chen *et al.*, 2023; Yosri *et al.*, 2024). Smart sensors detect specific biochemical and physicochemical changes that occur as food degrades. They can sense variations in temperature, humidity, gas concentration (such as oxygen, carbon dioxide, and ethylene), and microbial activity. These parameters are closely associated with food spoilage and shelf-life reduction (Duan *et al.*, 2025). Among these, gas sensors are highly significant as they can detect spoilage-related gases like ammonia and hydrogen sulfide, providing early warnings of microbial contamination or degradation (Chen *et al.*, 2023). A key aspect of enhanced shelf-life smart sensors is integrating intelligent packaging systems. These systems use indicators and sensors that respond to environmental changes, providing visible cues or transmitting data for automated quality control. TTIs track the cumulative time and temperature exposure of perishable products, enabling better inventory decisions. Other forms include RFID-based sensors and colorimetric sensors that change color in response to chemical changes, thus indicating the product's freshness (Chen *et al.*, 2023; Yosri *et al.*, 2024).

## 3.8 Nanobiosensors for Enhanced Sensitivity

Nanobiosensors represent an advanced analytical approach for improving sensitivity and specificity in food quality control. These nanoscale devices integrate biological recognition elements such as enzymes, antibodies, nucleic acids, or aptamers with physicochemical transducers to detect contaminants, pathogens, allergens, toxins, and spoilage markers at ultra-low concentrations (Yosri *et al.*, 2024). Their high surface-to-volume ratio, rapid response, and miniaturized design enable real-time, on-site monitoring with superior detection limits compared to conventional techniques. The operating principle involves selective binding between the biorecognition element and the target analyte, generating measurable signals via optical, electrochemical, piezoelectric, or thermal transduction (Chen *et al.*, 2023). The incorporation of NMs such as AuNPs, CNTs, quantum dots, graphene and MOFs significantly enhances sensor performance by increasing surface area for biomolecule immobilization and improving signal amplification through efficient electron transfer or fluorescence properties. Nanobiosensors have demonstrated high efficiency in detecting pathogens such as *E. coli*, *Salmonella* and *L. monocytogenes*, along with chemical contaminants including pesticides, antibiotics, heavy metals and mycotoxins (Chen *et al.*, 2024; Yosri *et al.*, 2024). For example, gold nanoparticle-based colorimetric sensors enable rapid visual detection of microbial contamination, while graphene-based electrochemical aptasensors offer picomolar sensitivity for antibiotic residues. Quantum dot-labeled immunosensors provide enhanced sensitivity for detecting trace aflatoxins in food matrices. Multiplexed nanobiosensors allow simultaneous detection of multiple analytes, supporting comprehensive food safety assessment. Integration with portable devices, wireless systems, and microfluidic lab-on-a-chip platforms has expanded their field applicability. However, challenges related to stability, reproducibility, scalability, and regulatory approval persist, necessitating further advancements in material design and biocompatibility for successful commercialization (Tutul *et al.*, 2023).

## 3.9 Nanomaterials and Surface Modifications

Nanomaterials and surface modification play a critical role in enhancing the performance of nanobiosensors for food packaging and shelf-life monitoring. These advancements improve sensitivity, selectivity and stability, enabling efficient detection of diverse analytes (Pandey *et al.*, 2025b). Nanomaterials such as NPs, nanowires and nanofilms exhibit unique physicochemical properties and a high surface-area-to-volume ratio, which enhance analyte interaction and amplify sensing signals, thereby improving overall biosensor efficiency. Surface modification is essential for the effective immobilization of biomolecules, including enzymes, antibodies, and nucleic acids, on to sensor surfaces (Mishra *et al.*, 2023). This facilitates specific target recognition and binding, ensuring accurate detection. Techniques such as self-assembled monolayers, chemical functionalization, and layer-by-layer deposition are widely employed to create stable and uniform bioactive interfaces, thereby increasing sensor sensitivity and specificity (Muthu *et al.*, 2025).

Bioconjugation strategies enable covalent attachment of biomolecules to NMs or modified surfaces, forming a robust sensing interface with high affinity toward target analytes. Nanocoatings and functionalization approaches enhance sensor durability, selectivity, and operational stability (Ramesh *et al.*, 2022; Gupta *et al.*, 2024). Quantum dots and fluorescent probes are frequently used as labeling agents due to their superior optical properties, including high signal-to-noise ratio and long-term fluorescence, enabling sensitive and multiplexed detection. Nanostructured surfaces developed through techniques such as nanopatterning and nanotexturing increase active sensing sites, improve analyte capture efficiency, and enhance binding specificity. These modifications also support biomolecule stability and reproducibility. Overall, the integration of advanced NMs and surface engineering significantly improves the analytical performance and reliability of nanobiosensors in food quality monitoring systems (Kakimova *et al.*, 2023; Gupta *et al.*, 2024).

## 3.10 Technical Challenges

The implementation of nanobiosensors for food packaging and shelf-life monitoring faces several critical technical challenges. A primary concern is ensuring sensor stability under variable environmental conditions, including fluctuations in temperature, humidity, mechanical stress, and exposure to complex food matrices during processing and storage. Sensor materials must maintain long-term functional integrity without compromising sensitivity or selectivity (Neethirajan *et al.*, 2018; Wang *et al.*, 2021b; Patial *et al.*, 2025). Advanced NMs such as graphene composites, MOFs and functionalized CNTs show promise due to their enhanced chemical and mechanical stability. Standardization and reproducibility also remain significant barriers. The lack of harmonized protocols for fabrication, functionalization and performance evaluation limits comparability and scalability. Establishing standardized methods for synthesis, characterization and validation is essential for regulatory approval and industrial adoption (Guruprasath *et al.*, 2024).

Improving sensitivity and selectivity for detecting trace-level contaminants such as pathogens, allergens, and chemical residues is another major challenge. This requires optimization of biorecognition elements (e.g. enzymes, antibodies, aptamers) and transducer interfaces to lower detection limits and minimize cross-reactivity. Emerging technologies, including CRISPR-based sensing, SPR and electrochemical nano-interfaces, offer enhanced specificity and enable detection at nanomolar to picomolar levels (Lei *et al.*, 2025). Reducing false positives and negatives in real-time applications is also critical. Integration of machine learning algorithms can improve signal interpretation, enhance predictive accuracy and adapt to dynamic food environments. Scalability and cost-effectiveness are crucial for commercialization. Techniques such as roll-to-roll printing, inkjet deposition and screen printing on flexible or biodegradable substrates can enable low-cost mass production and seamless integration into food packaging, supporting widespread industrial deployment (Rouf *et al.*, 2025).

## 3.11 Conclusion

Smart sensors and nanobiosensors have emerged as advanced tools for quality control in food, pharmaceutical, and environmental sectors, offering superior sensitivity, specificity, and real-time monitoring compared to conventional methods. This chapter outlines their principles, classifications, and recent advancements, highlighting their effectiveness in detecting contaminants, monitoring spoilage and ensuring product integrity. Nanotechnology-driven miniaturization has enhanced detection limits and response times, while functional NMs such as CNTs, graphene, quantum dots, and metal nanoparticles improve performance through high surface area, rapid electron transfer, and tunable surface properties. These features enable precise detection of biological and chemical markers, facilitating rapid decision making and reducing economic losses. Integration with IoT, wireless communication, and AI has enabled intelligent sensing systems capable of automated monitoring and predictive analysis, improving traceability and regulatory compliance. However, challenges such as stability, reproducibility, cost, scalability, and environmental safety of NMs remain critical concerns. Smart sensors and nanobiosensors represent a transformative, nondestructive, and highly accurate approach to quality monitoring. Continued interdisciplinary research is essential to address existing limitations and promote the widespread adoption for sustainable and reliable quality control systems.

# 4

# Risk Assessment of Nanomaterials in Food Applications

## 4.1 Introduction

Nanomaterials are increasingly employed in food applications to enhance quality, shelf life, safety, and nutritional value. However, the growing use of nanotechnology in the food sector has raised concerns about potential risks to human health and the environment. Risk assessment of NMs in food applications requires a thorough understanding of their unique physicochemical properties, exposure pathways, toxicokinetics, and potential adverse effects (Angelopoulou *et al.*, 2022; Gupta *et al.*, 2024a; Yadav *et al.*, 2025). Unlike bulk materials, NMs exhibit distinct behaviors due to their nanoscale size (1–100 nm), high surface-to-volume ratio, and increased reactivity. These characteristics influence their interactions with biological systems, potentially leading to unforeseen toxicological effects (Ahmad *et al.*, 2021; Hossain *et al.*, 2021; Chen *et al.*, 2023b). Common food-related NMs include nanosilver (AgNPs), titanium dioxide ($TiO_2$ NPs), zinc oxide (ZnO NPs), and silicon dioxide ($SiO_2$-NPs), among others (Chen *et al.*, 2023) (Table 4.1). They are used as antimicrobials, preservatives, colorants, carriers for bioactive compounds, and packaging enhancements (Awashra and Młynarz, 2023; Gupta *et al.*, 2024). The risk assessment process follows the conventional framework: hazard identification, dose–response assessment, exposure assessment, and risk characterization. Hazard identification for NMs involves *in vitro* and *in vivo* studies to evaluate their cytotoxicity, genotoxicity, oxidative stress potential, and inflammatory responses (Oomen *et al.*, 2018; Gupta *et al.*, 2024).

Several studies have demonstrated that NPs may translocate across the gastrointestinal tract and accumulate in organs, leading to systemic effects (Villamayor *et al.*, 2023; Gupta *et al.*, 2024). However, results are often inconsistent due to variations in particle size, shape, surface coating, and aggregation state. Exposure assessment for NMs in food is particularly challenging due to limited data on consumption patterns, NM concentrations in food matrices, and transformation during processing and digestion (Fig. 4.1; Chen *et al.*, 2022a; Yadav *et al.*, 2025). Current analytical techniques are often inadequate

**Table 4.1.** Common nanomaterials used in food applications and their associated risks based on current scientific evidence.

| Nanomaterial | Typical food/packaging uses | Main human exposure route(s) | Key risks/evidence from studies | Regulatory status / practical notes | References |
|---|---|---|---|---|---|
| Silver nanoparticles (AgNPs) | Antimicrobial coatings in food contact materials and packaging, sometimes proposed for food preservation or surface treatments | Oral (migration from packaging, residues on food), dermal/occupational | Antimicrobial activity can disturb gut microbiota; potential for ion-mediated toxicity; variable absorption but possible accumulation; genotoxicity and inflammatory responses reported in some *in-vivo/in-vitro* studies — risk depends strongly on particle size, coating, dose, and ion release | European Food Safety Authority (EFSA) concluded that the safety of AgNPs in some food contact uses must be assessed on a case-by-case basis; migration and release data are required. Use in food itself is tightly regulated | Ahmad *et al.*, 2021 |
| Titanium dioxide ($TiO_2$, E171) | Previously used as whitening/colorant in confectionery, bakery, sauces; also as a nanoparticulate in some coatings | Oral (direct additive or from coated foods) | EFSA found evidence raising concern for genotoxicity of $TiO_2$ particles; low oral absorption but potential for particle accumulation and DNA damage signals in some studies. Overall, EFSA (2021) concluded E171 can no longer be considered safe as a food additive | Following EFSA 2018 opinion, the EU moved to remove/limit E171 in foods; many manufacturers phased out E171 in foods. Regulatory stance is more restrictive in the EU; other jurisdictions vary | Blaznik *et al.*, 2021 |
| Silicon dioxide / Synthetic amorphous silica (E551) | Anti-caking agent in powdered foods, carriers for flavors; also used in some food contact materials | Oral (direct additive), migration from some materials | Most forms (SAS) historically considered low-toxicity, but concerns were raised about the presence of a nanofraction and potential for gastrointestinal interactions / uptake at nanoscale; EFSA re-evaluations highlight need for particle-size characterization and nanospecific data | EFSA re-evaluations concluded E551 acceptable at reported uses but recommended characterization because nanofractions may change hazard profile — case-by-case assessment | Fruijtier-Pölloth, 2016 |

| | | | | | |
|---|---|---|---|---|---|
| Zinc oxide nanoparticles (ZnO NPs) | Antimicrobial coatings, active packaging, antifungal/antibacterial surface treatments, possible postharvest coatings | Oral (migration/coating residues), inhalation (occupational) | ZnO NPs exhibit antibacterial properties but can generate reactive oxygen species (ROS); toxicity is highly dose- and formulation-dependent. In some models, ZnO can cause cytotoxicity, oxidative stress; but many food-application studies report low migration and low consumer exposure if properly embedded | Promising for packaging/antimicrobial use but requires migration testing and toxicological evaluation; recent reviews call for standardized testing and dose/exposure assessment | Sirelkhatim *et al.*, 2015; Blaznik *et al.*, 2021 |
| Lipid- and polymer-based nanocarriers (e.g. SLNs, NLCs, polymeric nanoparticles) | Nano-encapsulation of nutrients, flavors, bioactives (controlled release, improved bioavailability); edible nanofilms/coatings | Oral (intentionally ingested encapsulated ingredients) | Often made of food-grade lipids/polymers and tend to show lower inherent toxicity than inorganic NPs; main risks are from increased bioavailability of encapsulated actives (overdose risk), stability/transformation in GI tract, and potential for altered absorption or immune interactions. Toxicity is formulation-specific | Many formulations use GRAS/food-grade components; safety depends on composition, release profile, and the payload. Regulatory assessment focuses on both carrier and payload | Lu *et al.*, 2021; Tang *et al.*, 2023 |
| Nanocellulose (cellulose nanofibrils / nanocrystals) | Reinforcement for edible films, texture modifiers, stabilizers, packaging components | Oral (intended as ingredient/film), inhalation (dry powder during manufacture) | Generally low acute toxicity in many studies; main concerns: possible particle persistence, effects on gut mucosa or microbiome, and inhalation hazards for workers (fibrous forms). EFSA has been evaluating data gaps on digestion, persistence, and local effects | Emerging class with promising applications; EFSA projects and reviews call for more GI digestion and safety data before widespread food-use approval | Ong *et al.*, 2017; Xu *et al.*, 2024 |

*Continued*

**Table 4.1.** Continued

| Nanomaterial | Typical food/packaging uses | Main human exposure route(s) | Key risks/evidence from studies | Regulatory status / practical notes | References |
|---|---|---|---|---|---|
| Carbon nanotubes (CNTs) | Experimental: nanoreinforced packaging films for improved mechanical/barrier properties | Primarily dermal/inhalation (occupational)—oral possible if migration occurs | CNTs raise concerns because of high aspect ratio and persistence—some studies show pulmonary toxicity, fibrosis, and inflammation after inhalation; oral toxicity data are limited, but caution is advised due to similarity with other biopersistent fibrous particles | CNTs are generally not used inside food; occupational safety and migration testing are critical if used in packaging. Many regulators treat CNTs cautiously due to persistence and inhalation hazard evidence | Brito *et al.*, 2024; Papadiamantis *et al.*, 2024 |
| Gold nanoparticles (AuNPs) | Experimental uses: colorants, sensors, nano-encapsulates, and research uses in food authentication | Oral (intentionally or via migration) | AuNPs are often considered chemically inert, but biological effects depend on size, coating, and dose. Some *in vitro* studies show oxidative stress or uptake by gut cells at high doses; current evidence for consumer risk at realistic exposures is limited but not absent | Mainly experimental in foods; regulatory scrutiny focused on characterization, dose, and coatings | Gul *et al.*, 2025; Sati *et al.*, 2025 |

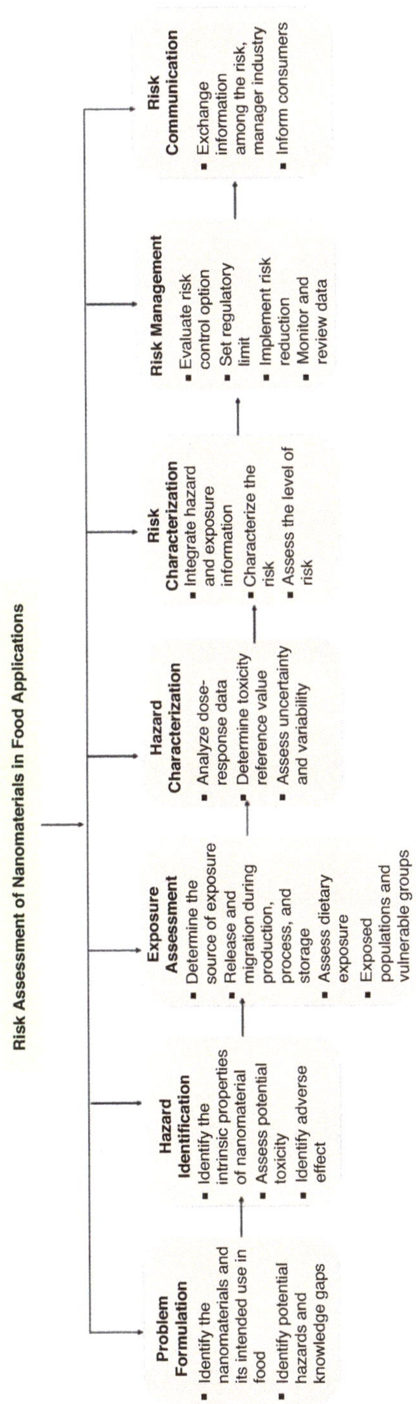


**Fig. 4.1.** Schematic representation of the risk assessment framework for nanomaterials in food applications.

for accurate detection and quantification of NMs in complex food systems. This hampers the ability to establish exposure thresholds and safe limits (Ghosh *et al.*, 2024). Another major concern is the lack of standardized protocols and regulatory guidelines for NMs. Existing toxicological models developed for bulk materials may not be suitable for assessing nanoscale risks (Chen *et al.*, 2022a). Regulatory agencies like EFSA, Food and Drug Administration (FDA), and FAO/World Health Organization (WHO) have initiated efforts to establish safety assessment frameworks for NMs in food, but comprehensive guidelines are still evolving (Ahmad *et al.*, 2021).

## 4.2 Sources of Exposure

**Direct additives:** Nano-encapsulated vitamins, antioxidants, antimicrobials, and colorants.
**Indirect exposure:** Migration of NPs from packaging materials into food.
**Environmental exposure:** Nanomaterials used in agriculture or processing may persist in soil and water, entering the food chain.

### 4.2.1 Direct additives

Nanomaterials are increasingly incorporated as food additives to enhance flavor, color, texture, shelf life, and nutritional value. Common examples include silver, $TiO_2$, silica, and ZnO NPs, which possess unique physicochemical properties such as high surface area, enhanced reactivity, and improved solubility (Biswas *et al.*, 2022). These attributes make them highly effective in food applications; however, their nanoscale nature raises concerns regarding potential risks to human health due to interactions at cellular and molecular levels (Sharma *et al.*, 2021b; Gupta *et al.*, 2024d). Human exposure to food-grade NMs occurs primarily through ingestion. Major sources include their intentional use as additives (e.g. colorants, preservatives, anticaking agents), migration from nanocomposite packaging, and contamination during processing (Table 4.2; Momirović *et al.*, 2023). Once ingested, NPs may be absorbed through the gastrointestinal tract and translocated to various organs. Their absorption and bioavailability depend on factors such as particle size, shape, surface charge, solubility, and interactions with food constituents (Momirović *et al.*, 2023). Risk assessment of NMs involves four key steps: hazard identification, hazard characterization, exposure assessment, and risk characterization. These steps evaluate toxicity, dose–response relationships, exposure levels, and overall health risks (Chen *et al.*, 2022). Potential toxic effects include oxidative stress, inflammation, genotoxicity, and disruption of gut microbiota. Despite progress in nanotoxicology, significant challenges remain, including a lack of standardized testing methods, limited long-term exposure data, and difficulties in detecting NMs in complex food matrices (Ashfaq *et al.*, 2023; Singh *et al.*, 2023b). Regulatory bodies emphasize case-by-case evaluation and recommend comprehensive characterization and safety assessment, often applying the precautionary principle where data are insufficient (Singh *et al.*, 2023; Gupta *et al.*, 2024).

**Table 4.2.** Various direct food additives containing nanomaterials, their functions, potential sources of exposure, and associated risks.

| Nanomaterial (examples) | Typical use (direct or indirect) | Function in food/ packaging | Potential exposure sources | Associated risks | References |
|---|---|---|---|---|---|
| Titanium dioxide ($TiO_2$, E171) | Direct food additive (whitening/gloss in confectionery, bakery, sauces) | Whitening, opacity, visual appeal; also used in some coatings. | Ingested in foods containing E171; residues in orally consumed products | EFSA concluded in 2021 that $TiO_2$ (E171) can no longer be considered safe as a food additive because a genotoxicity concern could not be ruled out (DNA strand breaks/ chromosomal damage reported in studies); additional indications of inflammation and possible immunotoxicity have been reported | Blaznik *et al.*, 2021 |
| Synthetic amorphous silica/ nanosilica (SAS, E551) | Direct food additive (anticaking agent in powdered foods, ingredients, supplements) | Prevents caking, improves powder flow; also used as a carrier in some supplements. | Ingested via powdered foods, spices, powdered infant formulas, supplements | EFSA re-evaluations and toxicology studies generally found no safety concern at reported uses/levels for food-grade SAS, but some *in vitro/in vivo* studies indicate possible effects at high exposures (gut epithelium, changes to gut barrier or inflammation in some models). Uncertainties remain over chronic low-dose effects and effects in sensitive groups | Fruijtier-Pölloth, 2016 |
| Silver nanoparticles (AgNPs) | Indirect — active antimicrobial agent incorporated into food contact materials / packaging; sometimes surface coatings | Antimicrobial (extend shelf life, reduce surface contamination) | Possible migration from packaging/ coatings into food, direct contact with treated surfaces, occupational exposure during manufacture | Migration studies give mixed results; there is concern about migration into food, silver systemic absorption, cytotoxicity, promotion of microbial resistance, and lack of robust long-term human data. Methodological variability in migration testing complicates risk estimates. | Ali *et al.*, 2023 |

*Continued*

**Table 4.2.** Continued

| Nanomaterial (examples) | Typical use (direct or indirect) | Function in food/ packaging | Potential exposure sources | Associated risks | References |
|---|---|---|---|---|---|
| Zinc oxide nanoparticles (ZnO NPs) | Indirect (food packaging), sometimes in active/coating films | Antimicrobial, UV barrier, improve mechanical and barrier properties of films | Possible migration from packaging; ingestion of food in contact with ZnO-containing films | ZnO NPs are widely studied for packaging; many studies find low migration under typical conditions but migration can occur depending on polymer, food type, and temperature. Toxicity is dose-dependent; at high doses ZnO NPs can produce oxidative stress and cytotoxicity in models – but recent reviews show promise for safe use when properly embedded and tested | Smaoui *et al.*, 2023 |
| Nanoclays / montmorillonite, layered silicates | Indirect (nanocomposite packaging films) | Improve mechanical strength, gas/ moisture barrier, and thermal stability of polymer films | Food packaged in nanoclay-reinforced films; potential nanoparticle release at seal/abrasion points | Improve barrier properties, but possible migration/release under certain conditions has been reported. Primary concerns: ingestion of liberated nanoparticles, limited long-term oral toxicity data, and the need for migration testing specific to the nanocomposite | Nath *et al.*, 2022; Deshmukh *et al.*, 2023 |

| | | | | | |
|---|---|---|---|---|---|
| Nanocellulose (CNF/CNC) | Indirect (bio-based packaging, edible films, stabilizers) | Reinforcement for films, barrier improvements, rheology modifier | Food contact with films; small amounts could be ingested if film is edible or degrades | Generally considered biocompatible and promising for sustainable films, but oral toxicity and particle fate depend on size/surface chemistry; regulatory and migration testing recommended before use in food contact. Recent reviews call for standardized safety testing | Pirozzi *et al.*, 2021; Jahangiri *et al.*, 2024 |
| Nano-encapsulation systems (liposomes, nano-emulsions, polymeric nanoparticles) | Direct (delivery vehicles for vitamins, antioxidants, flavors, probiotics, controlled-release ingredients) | Protect labile actives, increase solubility and bioavailability, control release, mask taste | Ingestion of fortified foods, beverages, supplements, and "smart foods" containing nanocarriers | Increased bioavailability can be beneficial (better delivery) but may also alter ADME (absorption, distribution, metabolism, excretion) and change toxicokinetics of the loaded compound; potential concerns include unexpected tissue distribution, altered allergenicity/immunogenicity, and limited long-term human safety data. Reviews emphasize case-by-case safety assessment | Ashfaq *et al.*, 2023 |
| Gold nanoparticles (AuNPs) and other metallic NPs (occasionally used) | Direct (rare), decorative colors or research ingredients; Indirect in sensors | Coloring, novelty/decorative uses; sensing in smart packaging (some research prototypes) | Rare ingestion from decorative foods (e.g. edible gold leaf, usually larger particles), experimental products | Metallic NPs may show size/shape-dependent toxicity *in vitro*; for most culinary uses (bulk gold leaf), particle sizes are large and inert, but engineered AuNPs used deliberately for nanoscale function require careful toxicology and migration assessment | Alberti *et al.*, 2021; Kusuma *et al.*, 2023 |

### 4.2.2 Indirect exposure

The integration of NMs in the food sector has gained significant attention due to their ability to improve food quality, safety, shelf life, and nutrient delivery. However, indirect exposure to NMs has emerged as a critical concern for human health (Ali *et al.*, 2023). Unlike direct incorporation, indirect exposure occurs unintentionally through food packaging, environmental contamination, agricultural inputs, or processing equipment. Nanocomposites used in packaging such as silver, $TiO_2$, and ZnO NPs can migrate into food during storage, posing potential risks (Kusuma *et al.*, 2023; Passeri *et al.*, 2024). Similarly, nanofertilizers and nanopesticides may accumulate in crops, entering the food chain. Environmental release of engineered NPs through industrial effluents and wastewater further contributes to bioaccumulation in aquatic organisms and trophic transfer. Nanocoated processing equipment may release particles due to abrasion or high temperatures (Blaznik *et al.*, 2021; Ali *et al.*, 2023). Risk assessment of indirect NM exposure is complex and depends on physicochemical properties such as size, shape, surface chemistry, and solubility. Standard risk assessment frameworks include hazard identification, dose–response evaluation, exposure assessment, and risk characterization. However, significant knowledge gaps remain, particularly regarding chronic low-dose exposure and nanoparticle behavior in the gastrointestinal system (Ashfaq *et al.*, 2023). Current regulatory frameworks lack standardized methods for evaluating the migration, bioavailability, and toxicity of indirectly used NMs. Therefore, advanced nanosafety strategies such as predictive toxicology, *in vitro* gastrointestinal models, and probabilistic exposure assessments are essential. International organizations emphasize improved risk governance, transparency, and labeling (Ashfaq *et al.*, 2023; Kusuma *et al.*, 2023).

### 4.2.3 Environmental exposure

Nanomaterials are increasingly utilized in food systems due to their antimicrobial properties, ability to extend shelf life, and enhanced nutrient delivery. Despite these benefits, their integration raises concerns regarding potential human health and environmental risks. Environmental exposure to NMs occurs throughout their life cycle, including production, application, disposal, and degradation, leading to their presence in air, water, soil, and biota (Ashfaq *et al.*, 2023; Smaoui *et al.*, 2023). Consequently, human exposure may occur via ingestion, inhalation, or dermal contact. In food systems, direct exposure arises from their intentional use as additives, packaging materials, or processing aids, such as AgNPs for antimicrobial packaging and nanosilica or nano-$TiO_2$ as functional additives. Indirect exposure may occur through environmental contamination, resulting in bioaccumulation in crops and livestock (Ashfaq *et al.*, 2023; Chen *et al.*, 2023). Occupational exposure is also significant during manufacturing and handling processes. Risk assessment of NMs requires evaluation of toxicity, bioavailability, and persistence. Owing to

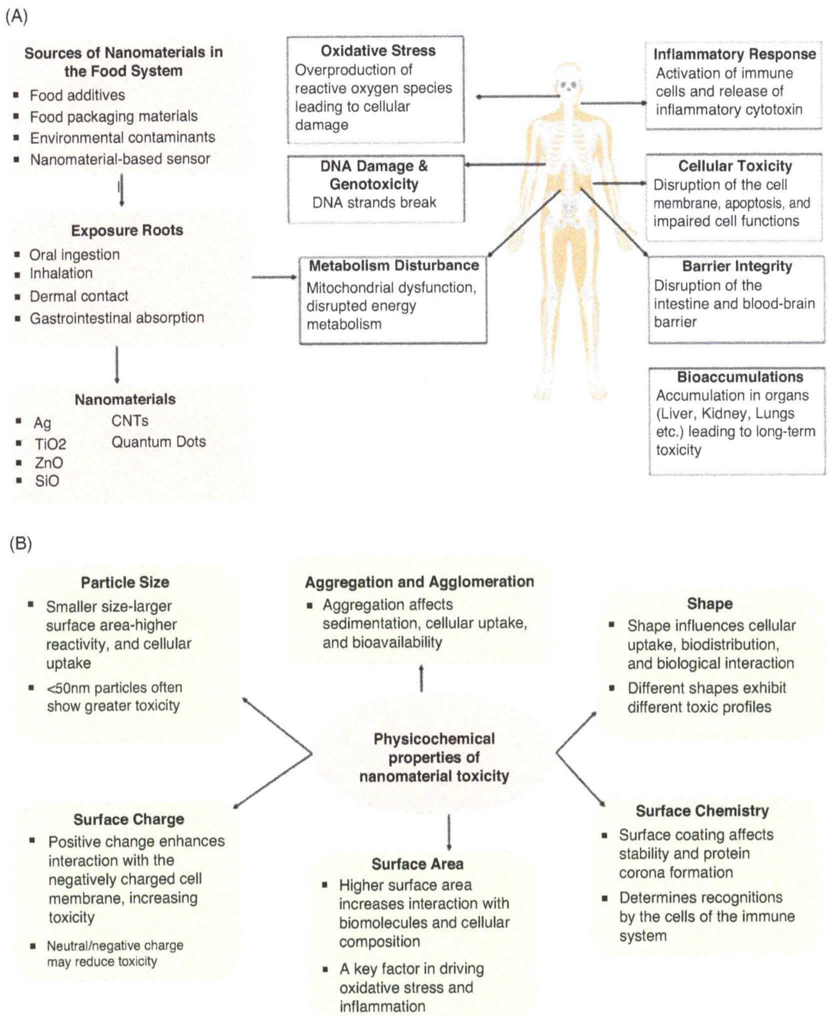


**Fig. 4.2.** (A) Toxicological effects of nanomaterials in food systems, showing key biological responses including oxidative stress, (B) Physicochemical determinants influencing nanomaterial toxicity.

their high surface area and reactivity, NMs can induce oxidative stress, inflammation, and cellular damage, though their effects depend on physicochemical properties such as size, shape, and surface characteristics (Ashfaq *et al.*, 2023; Chen *et al.*, 2023). Conventional risk assessment approaches must be adapted to address nanospecific behaviors, necessitating standardized methods for detection and characterization in complex matrices. Comprehensive risk management should incorporate life cycle analysis and environmental fate modeling to assess long-term impacts. Regulatory frameworks are evolving, emphasizing precautionary principles and safety evaluation. Ensuring transparency, interdisciplinary collaboration, and robust scientific assessment is essential to balance technological benefits with health and environmental protection (Gupta *et al.*, 2024; Jahangiri *et al.*, 2024).

## 4.3 Toxicological Concerns

Nanomaterials are increasingly incorporated into food products and packaging due to their unique physicochemical properties, including high surface area, reactivity, and size-dependent functionalities (Chen *et al.*, 2023). However, their interaction with biological systems raises toxicological concerns necessitating a comprehensive risk assessment (Fig. 4.2A; Abouzeinab *et al.*, 2023). Upon ingestion, NMs may translocate across gastrointestinal barriers, interact with gut microbiota, and accumulate in organs, potentially inducing oxidative stress, inflammation, and genotoxicity (Ahmad *et al.*, 2023; Jahangiri *et al.*, 2024). The risk assessment process involves hazard identification, dose–response assessment, exposure assessment, and risk characterization, yet these stages are challenged by the lack of standardized methodologies, inadequate data on chronic exposure and diverse NM compositions (Table 4.3; Chen *et al.*, 2023; Jahangiri *et al.*, 2024). NM behavior may be influenced by the food matrix, digestion dynamics, and surface modifications, complicating their toxicokinetic profiling (Ahmad *et al.*, 2021, 2023). Advanced *in vitro* and *in vivo* models, along with omics approaches and physiologically based pharmacokinetic (PBPK) models, are critical in elucidating their biological fate. Regulatory agencies emphasize the need for case-by-case evaluation of NMs, incorporating physicochemical characterization, bioavailability, and toxicity endpoints. Bridging knowledge gaps through interdisciplinary research is essential to ensure consumer safety and foster responsible innovation in food nanotechnology (Abouzeinab *et al.*, 2023; Ahmad *et al.*, 2023).

### 4.3.1 Size

Nanomaterials are increasingly applied in food systems due to their unique physicochemical properties, particularly their nanoscale size and high surface reactivity, which enhance shelf life, bioavailability, and sensory quality. However, these same properties raise concerns regarding potential health risks. The nanoscale range (1–100 nm) significantly increases the surface area-to-volume ratio, leading to greater reactivity and enhanced interaction with biological systems. As a result, NMs may induce oxidative stress, inflammation and genotoxic effects upon ingestion (Gao and Lowry, 2018; Lopez *et al.*, 2022). Particle size is a critical determinant of NM toxicity. Smaller NPs generally exhibit higher toxicity than larger counterparts due to increased surface reactivity, dissolution rate, and bioavailability (Lopez *et al.*, 2022). Graphene oxide NPs have demonstrated size-dependent

**Table 4.3.** Risk assessment of nanomaterials in food applications.

| Key parameter | Description | Importance of risk assessment | Example/consideration | Reference |
|---|---|---|---|---|
| Physicochemical properties | Includes particle size, shape, surface charge, crystalline structure, solubility, agglomeration state, and surface chemistry. | Determines nanomaterial reactivity, mobility, and interaction with biological systems. | Smaller nanoparticles (<100 nm) may penetrate cell membranes more easily than larger ones. | Gatoo *et al.*, 2014; Khan *et al.*, 2022 |
| Stability in food matrix | Behavior of nanomaterials in various food matrices (pH, fat, protein, carbohydrate content). | Affects exposure level and bioavailability during digestion. | Nanosilver in dairy may behave differently than in acidic beverages. | Bouwmeester *et al.*, 2009; Lopez *et al.*, 2022 |
| Digestive transformation | Fate of nanomaterials during gastrointestinal passage (e.g. dissolution, aggregation, interaction with enzymes). | Determines systemic exposure and potential toxicity. | $TiO_2$ nanoparticles may undergo partial dissolution in stomach acid. | Ryabtseva *et al.*, 2023 |
| Bioavailability and absorption | The extent to which nanomaterials cross intestinal barriers and enter the bloodstream. | Critical for systemic exposure and organ distribution. | Carbon nanotubes may accumulate in the liver and spleen. | Xuan *et al.*, 2023 |
| Toxicokinetics | Absorption, distribution, metabolism, and excretion (ADME) of nanomaterials. | Determines persistence, accumulation, and clearance from the body. | Nano-ZnO absorbed and excreted differently compared to bulk ZnO. | Choi and Choy, 2014; Thu *et al.*, 2023 |
| Toxicological endpoints | Includes cytotoxicity, genotoxicity, immunotoxicity, reproductive and developmental toxicity. | Essential for hazard characterization. | AgNPs can induce oxidative stress and DNA damage. | Dusinska *et al.*, 2017; Ferdous and Nemmar, 2020 |

*Continued*

**Table 4.3.** Continued

| Key parameter | Description | Importance of risk assessment | Example/consideration | Reference |
|---|---|---|---|---|
| Exposure assessment | Estimation of consumer exposure levels via food, packaging, and additives. | Helps identify risk level based on intake. | Migration of nanoclays from packaging into food oils. | de Oliveira Mallia *et al.*, 2022 |
| Consumer susceptibility | Differences in vulnerability (children, elderly, immunocompromised individuals). | Identifies high-risk populations. | Infants may be more sensitive to nanoparticles in formula. | Hofer *et al.*, 2022; Fagbohun *et al.*, 2025 |
| Environmental fate and carryover | Release, persistence, and bioaccumulation of nanomaterials in ecosystems. | Long-term indirect risk via food chain. | Nano-$TiO_2$ accumulation in aquatic organisms leading to biomagnification. | Li *et al.*, 2022 |
| Analytical detection and characterization methods | Techniques like Transmission Electron Microscope (TEM), Scanning Electron Microscope (SEM), Dynamic Light Scattering (DLS), Inductively Coupled Plasma Mass Spectrometry (ICP-MS), Single Particle Inductively Coupled Plasma Mass Spectrometry (sp-ICP-MS), and Nuclear Magnetic Resonance (NMR). | Accurate detection in complex food matrices is key for risk evaluation. | Difficulty in detecting nanoparticles in processed foods. | Tiede *et al.*, 2008; Gao and Lowry, 2018 |

toxicity, with smaller particles showing greater adverse effects in embryonic zebrafish, primarily through oxidative stress mechanisms (Li *et al.*, 2022; Lopez *et al.*, 2022). Smaller AgNPs are more likely to induce organ damage owing to their enhanced toxicokinetics and higher ion release (Al-Doaiss *et al.*, 2020; Li *et al.*, 2022). Zinc oxide NMs also exhibit size-dependent effects, where smaller particles induce significant genotoxicity in human lymphocytes, while larger particles tend to cause cytotoxicity (Sahani and Sharma, 2021; Li *et al.*, 2022). Reduced particle size enhances the ability of NMs to cross biological barriers and interact with cellular components, thereby increasing their toxic potential and necessitating careful risk assessment in food applications (de Oliveira Mallia *et al.*, 2022).

### 4.3.2 Shape and aspect ratio

The shape and aspect ratio of NMs are critical factors that significantly influence their toxicity. These characteristics determine how NMs interact with biological systems, leading to varying degrees of cellular uptake, inflammation, and overall toxicity (de Oliveira Mallia *et al.*, 2022; Berthing *et al.*, 2023). An earlier study reported that, when long, needle-like structures such as CNTs and asbestos fibers can activate the NLRP3 inflammasome, triggering a pro-inflammatory response in human macrophages. This indicates that high aspect ratio NMs can induce inflammation through mechanisms similar to known hazardous materials (de Oliveira Mallia *et al.*, 2022). Gallium phosphide nanowires with a high aspect ratio have been shown to cause genotoxicity and acute inflammation in lung tissues of mice, highlighting the role of shape and aspect ratio in pulmonary toxicity (Berthing *et al.*, 2023). In another study, polyaniline NMs with lower aspect ratios were found to be more cytotoxic, inducing greater necrosis and reactive oxygen species (ROS) production in human fibroblast cells compared to their higher aspect ratio counterparts (Berthing *et al.*, 2023). Cellulose NMs with a high aspect ratio, such as nanocrystals, are more likely to cause pro-inflammatory effects and immune system activation compared to nanofibrils, which have a lower aspect ratio (Fig. 4.2.B; Wang *et al.*, 2019; Ahmadi *et al.*, 2024). Studies on AgNPs have shown that particles with lower aspect ratios and higher circularity tend to be more toxic at both lethal and sublethal levels (Auclair and Gagne, 2022). These findings underscore the importance of considering the shape and aspect ratio of NMs in toxicity assessments, as they play a crucial role in determining the biological outcomes of NM exposure.

### 4.3.3 Dose and structure

The toxicity of NMs is significantly influenced by both dose and structural properties. These factors determine how NMs interact with biological systems, affecting their distribution, cellular uptake, and potential to cause harm. A recent study suggested that the toxicity of mesoporous polydopamine nanoparticles (MPDA) is dose-dependent. The findings indicated that administering MPDA at low (3.61 mg/kg) and medium doses (10.87 mg/kg) did not significantly impact body weight, organ index, or routine blood parameters. However, injecting a high dose of MPDA (78.57 mg/kg) resulted in disruptions to the gut microbiota, activation of inflammatory pathways due to abnormal metabolism of bile acids and unsaturated fatty acids, and potential oxidative stress injury (Chen *et al.*, 2023; Awasthi *et al.*, 2024). In an investigation of two-dimensional $TiO_2$ nanosheets, a dose of 10 mg/kg resulted in the gradual accumulation of the NPs in the liver and spleen, with slight abnormalities in liver biochemistry observed over 30 days (Song *et al.*, 2014; Awasthi *et al.*, 2024). Previous study also demonstrated that ZnO NPs caused developmental neurotoxicity in a dose-dependent manner (0.1, 1, 10, 50, and 100 µg/ml), with high doses triggering Parkinson's disease-like symptoms through oxidative stress mechanisms (Jin *et al.*, 2021). The combination of dose and structural factors can also be crucial in determining toxicity. For example, a study on ZnO NMs revealed that smaller-sized ZnO nanorods induced more severe toxic effects, particularly at higher doses, leading to oxidative stress and mitochondrial damage (Jin *et al.*, 2021; Awasthi *et al.*, 2024).

### 4.3.4 Cellular uptake

Cellular uptake is a critical factor in the risk assessment of NMs used in food applications. The interaction of NMs with biological systems is highly dependent on their physicochemical properties, including size, shape, surface charge, solubility, and coating (Awasthi *et al.*, 2024). Nanomaterials may be absorbed across the gastrointestinal epithelium upon ingestion via passive diffusion, endocytosis, or transcytosis. Once internalized, they can interact with organelles and biomolecules, potentially leading to cytotoxicity, oxidative stress, and inflammation. The ability of NMs to cross biological barriers and accumulate in secondary organs such as the liver, spleen, and brain raises further concerns (Jin *et al.*, 2021; Medina-Ramirez *et al.*, 2024). Risk assessment must account for these cellular-level interactions to determine potential adverse effects. *In vitro* models such as $CaCO_2$ and Human Tumour 29 - Methotrexate-adapted (HT29-MTX) cells are widely used to simulate intestinal uptake. However, discrepancies between *in vitro* and *in vivo* responses highlight the need for more physiologically relevant models (Gibbs *et al.*, 2021). Nanomaterials may transform the gastrointestinal tract, including dissolution, agglomeration, or interaction with food components, altering their bioavailability and toxicity. Therefore, risk characterization requires a holistic approach integrating cellular uptake data, biotransformation, dose–response relationships, and exposure scenarios (Gupta *et al.*, 2024).

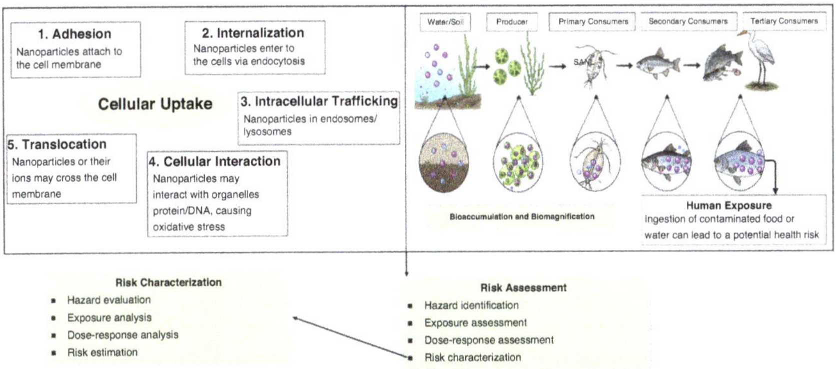


**Fig. 4.3.** Cellular uptake of nanoparticles, their accumulation in the food chain, and the step-by-step process used to assess and characterize their potential health risks.

### 4.3.5 Bioaccumulation potential

The increasing integration of nanotechnology in food processing, packaging and preservation, the bioaccumulation potential and risk assessment of NMs in food applications are crucial aspects of nanotoxicology (Chen *et al.*, 2023). Nanomaterials, due to their small size and high surface area, can interact uniquely with biological systems. Once ingested, they may cross biological membranes, accumulate in organs, and cause potential toxic effects (Fig. 4.3) (Gupta *et al.*, 2024). The bioaccumulation of NMs depends on several factors, including particle size, shape, surface charge, solubility, and surface functionalization (Simbine *et al.*, 2019; Medina-Ramirez *et al.*, 2024). Chronic exposure through dietary intake raises concerns regarding their persistence and potential for trophic transfer in the food chain. Risk assessment involves a comprehensive evaluation of hazard identification, dose–response relationships, exposure assessment, and risk characterization (Ahmadi *et al.*, 2024; Fujihara and Nishimoto, 2024). The lack of standardized testing protocols and the complex behavior of NMs in biological matrices make risk prediction challenging. Several studies have shown that metal-based NMs like AgNPs, $TiO_2$NPs and ZnONPs can accumulate in liver, kidney, and intestinal tissues, potentially leading to oxidative stress, inflammation, and genotoxic effects (Simbine *et al.*, 2019; Fujihara and Nishimoto, 2024). Regulatory agencies have emphasized the need for developing robust *in vitro* and *in vivo* models to assess bioavailability, metabolism, and long-term toxicity of these materials. Nanomaterial characterization under realistic food matrices and conditions is also essential for accurate risk assessment. Integration of omics tools, predictive modeling, and life cycle analysis can provide a more holistic understanding of NM behavior and aid in safer food nanotechnology development (Ahmadi *et al.*, 2024; Medina-Ramirez *et al.*, 2024).

## 4.4 Risk Assessment Framework

The increasing integration of nanotechnology in food systems has raised significant interest and concern over the safety of NMs used in various applications, including food processing, packaging, preservation, and fortification (Simbine *et al.*, 2019; Hossain *et al.*, 2021). Owing to their unique physicochemical properties, such as nanoscale size, high surface area-to-volume ratio, and enhanced reactivity, NMs interact with biological systems in ways that may not be predictable from their bulk counterparts (Fig. 4.3; Alfei, 2020; Fujihara and Nishimoto, 2024). Therefore, a structured and comprehensive risk assessment framework is essential to ensure the safe use of NMs in food applications. Risk assessment of NMs generally follows the classical four-step model: hazard identification, hazard characterization, exposure assessment, and risk characterization (EFSA Scientific Committee, 2018). However, the distinct nature of NMs necessitates the adaptation of conventional methodologies. For instance, the first step involves determining whether an NM poses a potential health hazard based on its composition, morphology, surface charge, and functionalization (Takeuchi *et al.*, 2014). This is followed by hazard characterization, which quantifies the adverse effects and determines dose–response relationships. Toxicological evaluations, such as *in vitro* cytotoxicity, genotoxicity, and *in vivo* studies in animal models, are essential to understand acute and chronic effects.

Exposure assessment involves identifying the routes through which consumers are exposed to NMs, typically oral ingestion, dermal contact, or inhalation, and quantifying the levels of exposure based on concentration in food products and consumption patterns. Risk characterization integrates hazard and exposure data to estimate the likelihood and severity of adverse effects, supporting regulatory decision making (EFSA Scientific Committee *et al.*, 2018). An essential challenge in the risk assessment of NMs is the lack of standardized protocols and validated testing methods. Moreover, NMs may undergo transformations in the food matrix or gastrointestinal tract, affecting their bioavailability and toxicity (Alfei, 2020; Yilma *et al.*, 2024). To address these issues, a multidisciplinary approach involving materials science, toxicology, food chemistry, and regulatory science is necessary. Several regulatory agencies, including EFSA, FDA, and International Organization for Standardization (ISO), have provided guidelines for NM risk assessment, emphasizing the need for case-by-case evaluations (EFSA Scientific Committee *et al.*, 2018; Yilma *et al.*, 2024).

### 4.4.1 Dose–response assessment

Nanomaterials in food applications, whether used in packaging, processing, or as additives, pose emerging concerns regarding their potential toxicity and associated risks to human health (Wang *et al.*, 2021a). Dose–response assessment is a critical component of NM risk evaluation, determining the relationship between the extent of exposure and the probability of adverse health outcomes. Owing to their nanoscale dimensions (<100 nm), NMs exhibit unique physicochemical properties such as high surface area, reactivity, and altered bioavailability, which may influence their interaction with biological systems (Usmani *et al.*, 2024). Traditional toxicological models may not fully capture the dose–response relationships of these materials. NMs may exhibit nonlinear or threshold-independent responses, especially when they translocate across biological barriers or induce oxidative stress, inflammation, or genotoxicity even at low doses (Alfei, 2020; Wang *et al.*, 2021). A key challenge in the dose–response assessment is the lack of standardized metrics for dose quantification (e.g. mass, number, surface area) and the variability in their physicochemical attributes, including size, shape, surface charge, and solubility (Wang *et al.*, 2021; Usmani *et al.*, 2024). Exposure routes such as oral ingestion through food additives, leaching from nanopackaging, or indirect contamination add to the complexity. Studies have indicated that certain NMs, such as AgNPs, $TiO_2$NPs, and ZnONPs, may accumulate in tissues or interact with gut microbiota, potentially altering metabolism and immune responses (Simbine *et al.*, 2019; Wang *et al.*, 2021; Fujihara and Nishimoto, 2024). Regulatory frameworks and risk assessment protocols must integrate nanospecific endpoints and dose metrics to improve the accuracy of predictions. *In vitro* and *in vivo* studies, along with PBPK models, are vital for understanding the biological fate and dose thresholds. A precautionary approach is recommended where uncertainties remain high (Tschiche *et al.*, 2022; Fujihara and Nishimoto, 2024).

### 4.4.2 Risk characterization

Risk characterization of NMs in food involves the integration of hazard identification, dose–response assessment, exposure assessment, and risk estimation, providing a comprehensive understanding of the likelihood and severity of adverse effects resulting from their use (Fig. 4.3; Alfei, 2020; Usmani *et al.*, 2024). One major challenge in characterizing the risks associated with NMs is their complex behavior in biological systems, including potential for bioaccumulation, persistence, and interactions at the molecular or cellular level (Tschiche *et al.*, 2022; Fujihara and Nishimoto, 2024). Unlike conventional substances, NMs may translocate across biological barriers such as the gastrointestinal tract, blood–brain barrier, and placenta. Their small size and surface modifications often influence their toxicokinetics, complicating traditional dose–response models (Wang *et al.*, 2021; Rizvi *et al.*, 2022). Therefore, the risk assessment process must account for NP-specific parameters such as size, shape, surface charge, solubility, and coating material. Toxicological studies suggest that some NMs, such as AgNPs, $TiO_2$NPs, and ZnONPs, exhibit cytotoxic, genotoxic, pro-inflammatory, and oxidative stress-related effects, depending on their concentration and exposure duration (Table 4.3; Wang *et al.*, 2021a; Fujihara and Nishimoto, 2024). Moreover, the long-term effects of chronic low-dose exposure through the food chain remain largely unexplored (Usmani *et al.*, 2024). The potential for transformation of NMs during food processing or digestion, resulting in altered bio-availability or toxicity, adds further complexity to risk characterization. Risk characterization also involves exposure assessment, which estimates the extent of human intake of NMs via food, water, or inhalation during production and handling (Usmani *et al.*, 2024). However, reliable exposure data are scarce due to the lack of mandatory labeling, analytical detection challenges, and variations in NM incorporation across food matrices. Probabilistic and deterministic models, along with *in vitro* and *in vivo* studies, are employed to support exposure estimation and help identify at-risk populations such as children, pregnant women, and immunocompromised individuals (Verleysen *et al.*, 2021; Rizvi *et al.*, 2022; Fujihara and Nishimoto, 2024).

## 4.5 Conclusion

The integration of NMs in food applications, ranging from food packaging and preservation to nutrient delivery and quality enhancement, has shown remarkable potential to revolutionize the food industry. However, these innovations come with critical safety concerns that necessitate a comprehensive risk assessment. This chapter underscores the multifaceted nature of risk assessment of NMs, involving physicochemical characterization, exposure evaluation, toxicological profiling, and regulatory oversight. NMs exhibit unique properties such as high surface area-to-volume ratios, reactivity, and the ability to interact at the molecular level, which can not only enhance functional benefits but also pose unforeseen risks to human health and the environment. Current toxicological studies indicate that some engineered NMs may cause cytotoxicity, genotoxicity, oxidative stress, or bioaccumulation, particularly when ingested through food. The complexity increases due to the lack of standardized testing protocols, insufficient long-term exposure data, and limited understanding of nanoparticle interactions within the gastrointestinal tract and food matrices.

Risk assessment frameworks must evolve to include nanospecific considerations, including novel *in vitro* and *in vivo* models, high-throughput screening methods, and advanced computational modeling to predict behavior, bioavailability, and toxicity. Regulatory agencies worldwide, such as EFSA and FDA, have begun incorporating NM-specific guidelines, but a harmonized international framework remains essential. Nanotechnology presents significant advantages for food innovation, its safe application depends on robust, science-based risk assessment methodologies. Interdisciplinary collaboration among material scientists, toxicologists, food technologists, and regulators is vital for developing accurate risk profiles and ensuring consumer protection. A precautionary yet progressive approach is necessary, embracing technological advances while continuously updating safety standards based on emerging scientific evidence. Only through rigorous and transparent risk assessment can NMs be responsibly and sustainably integrated into the global food system.

# 5 Nanotechnology in Addressing the Climate Impact of Food Systems

## 5.1 Introduction

Nanotechnology is emerging as a powerful tool in addressing the climate impact of food systems, which are responsible for nearly one-third of global GHG emissions. Food production, processing, packaging, transportation, and waste disposal all contribute to environmental degradation and climate change (Elrashidi *et al.*, 2022; Elzein, 2024). By integrating nanotechnology into various stages of the food supply chain, it is possible to enhance efficiency, reduce resource use, and mitigate emissions, offering a pathway toward sustainable and climate-resilient food systems. One of the key areas where nanotechnology contributes is in sustainable agriculture. Nanofertilizers and nanopesticides enable precision farming by delivering nutrients and active compounds in a controlled and targeted manner, thereby reducing the overuse of agrochemicals and minimizing leaching into the environment (Mittal *et al.*, 2020; Shoukat *et al.*, 2024). This precision not only enhances crop yield but also lowers the carbon footprint of agricultural inputs. Nanocarriers, such as liposomes, dendrimers, and polymeric NPs, can encapsulate fertilizers and pesticides, thereby improving their stability and efficacy under variable environmental conditions (Musa *et al.*, 2025).

Nanotechnology also supports climate-smart food processing. Nanoencapsulation of bioactive compounds, antioxidants, and antimicrobial agents enhances shelf life, minimizes spoilage, and preserves food quality, thereby reducing food loss and waste, which is a notable contributor to GHG emissions (Elzein, 2024). Nanosensors in packaging can monitor temperature, humidity, and microbial contamination in real time, ensuring food safety and reducing the need for conservative overpackaging or excess energy use in storage (Guleria *et al.*, 2023). In the domain of packaging, NMs such as nanoclay, nanosilver, and CNTs are being used to develop biodegradable, antimicrobial, and gas-barrier packaging materials (Ashfaq *et al.*, 2022). These innovations reduce dependence on petroleum-based plastics and prolong the shelf life of perishable products, further helping to lower emissions associated with food waste. Additionally, nanostructured membranes and filters are being used in water purification and food processing to reduce energy use and improve process efficiency (Shoukat *et al.*, 2024).

Nanotechnology also offers potential in the development of climate-resilient crop varieties (Shoukat *et al.*, 2024). Nanoparticles can facilitate gene delivery and genome editing, aiding in the development of crops with enhanced tolerance to drought, salinity, and temperature fluctuations, traits that are critical under climate change scenarios (Shoukat *et al.*, 2024). Moreover, nano-enabled soil amendments and biosensors can monitor soil health, moisture, and nutrient status in real time, thereby supporting adaptive and efficient farming practices (Table 5.1; Singh *et al.*, 2024). The application of nanotechnology in food systems requires careful assessment of potential environmental and health risks. The fate, transport, and toxicity of NPs in soil, water, and food chains must be thoroughly understood to ensure safe deployment. Regulatory frameworks and consumer awareness must evolve alongside technological advancements to ensure public trust and sustainable integration of nanotechnology into food systems (Fig. 5.1; Scott *et al.*, 2018; Shah *et al.*, 2024).

## 5.2 Applications of Nanotechnology in Food Systems for Climate Mitigation

Nanotechnology has emerged as a transformative approach in food systems, offering innovative solutions to enhance sustainability and contribute to climate change mitigation (Biswas *et al.*, 2022; Apoorva *et al.*, 2023). The integration of nanotechnology into food production, processing, packaging, and distribution can reduce GHG emissions, improve resource efficiency, and minimize food waste, thereby supporting global efforts to curb climate impacts (Chausali *et al.*, 2023; Shoukat *et al.*, 2024). At the production level, nanotechnology enhances precision agriculture by enabling targeted delivery of nutrients, pesticides, and water, reducing excessive use and associated emissions (Chausali *et al.*, 2023). Nanofertilizers and nanopesticides exhibit higher efficiency and controlled release properties, reducing leaching and volatilization losses that contribute to GHG emissions. For example, encapsulation techniques using NPs allow slow release of agrochemicals, decreasing nitrous oxide emissions from fertilizers, a potent GHG (Shoukat *et al.*, 2024).

In food processing, nanoscale materials improve energy efficiency by optimizing heat and mass transfer. Nano-encapsulation of bioactive compounds preserves nutritional quality while reducing processing time and energy consumption. Nanomaterials can be used to develop novel enzymes and catalysts that facilitate greener and less energy-intensive food processing technologies (Chausali *et al.*, 2023; Shah *et al.*, 2024). This reduction in energy demand lowers the carbon footprint of food products. Nanotechnology also revolutionizes food packaging through the development of active and intelligent packaging materials (Shah and Krishnakumar, 2025). Nano-enabled packaging extends shelf life by providing improved barriers against oxygen, moisture,

**Table 5.1.** Applications of nanotechnology across the food system and their impact on climate mitigation.

| Food system stage | Nanotechnology applications | Climate mitigation impact | Examples/details | References |
|---|---|---|---|---|
| Agriculture (Production) | Nanofertilizers and nanopesticides | Reduce excessive use of chemical fertilizers/pesticides, lower GHG emissions from fertilizer production and application | Controlled release of nutrients, improved uptake efficiency, and reduction in nitrous oxide emissions | Mathur *et al.*, 2022; Rana *et al.*, 2024 |
| | Nanobiosensors for soil and crop monitoring | Precision farming reduces resource wastage and energy use | Real-time detection of soil nutrients, water stress, and pathogens | Sarabandi *et al.*, 2025 |
| | Nanocoatings for seeds | Enhanced germination, stress tolerance, reduced input requirements | Improved drought and salinity resistance | Singh *et al.*, 2024 |
| Food processing | Nanomaterials for energy-efficient food preservation | Lower energy consumption in cooling, freezing, and storage | Nano-encapsulation of antimicrobials to extend shelf life | Chaudhary and Janmeda, 2025 |
| | Nanostructured catalysts for green processing | Reduced energy use and emissions in food manufacturing | Enzymatic activity enhancement, cleaner production | Biswas *et al.*, 2022 |
| Food packaging | Nanocomposite packaging materials | Reduce food waste by extending shelf life, lowering methane emissions from landfills | Silver nanoparticles, clay nanocomposites for antimicrobial/oxygen barrier | del Rosario Herrera-Rivera *et al.*, 2024 |
| | Active and intelligent packaging with nanosensors | Reduce spoilage and optimize logistics, minimizing food loss | Time–temperature indicators, freshness sensors | Ahari and Soufiani, 2021 |

**Table 5.1.** Continued

| Food system stage | Nanotechnology applications | Climate mitigation impact | Examples/details | References |
|---|---|---|---|---|
| Food distribution and supply chain | Nano-enabled cold storage and logistics | Reduce energy use in transport and refrigeration | Nanocoatings for insulation, phase change materials | Peng *et al.*, 2025 |
| | Blockchain–nanosensor integration | Traceability and reduced waste along supply chain | Early spoilage detection, optimized routing | Tunc and Akyuz, 2025 |
| Consumption | Nano-encapsulation for nutrient delivery | Improves food quality and reduces need for resource-intensive fortification | Vitamins, antioxidants are delivered efficiently | Altemimi *et al.*, 2024 |
| | Nanobased water purification for food preparation | Safe water reduces dependence on bottled water and cuts plastic waste | Nanofiltration membranes, carbon nanotubes, nano-$TiO_2$ | Rabiee *et al.*, 2023 |
| Waste management and circular economy | Nanocatalysts for bioenergy generation from food waste | Enhanced biogas/biohydrogen yields, reducing methane emissions | Nano-Fe, nano-ZnO in anaerobic digestion | Kumar *et al.*, 2025 |
| | Nano-enabled composting and nutrient recovery | Recycling nutrients back into soil, reducing synthetic fertilizer use | Nanostructured biochar, metal oxide nanoparticles | Ferdeş *et al.*, 2022 |

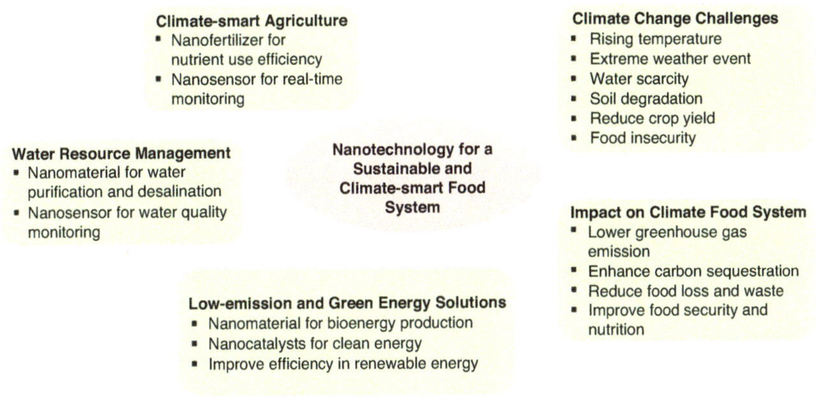


**Fig. 5.1.** Nanotechnology in addressing the climate change of the food system.

and microbial contamination. This leads to a significant decrease in food spoilage and waste, a major source of methane emissions from landfills. Intelligent packaging embedded with nanosensors monitors food freshness and quality in real time, helping to reduce unnecessary disposal and optimize supply chains (Bandyopadhyay and Ray, 2023; Shah and Krishnakumar, 2025). Moreover, nanotechnology contributes to alternative protein production, which is critical for climate mitigation. Nanostructured scaffolds and delivery systems are being developed for cultured meat and plant-based protein alternatives, enhancing texture, flavor, and nutrient delivery, making these options more acceptable to consumers and potentially reducing reliance on conventional livestock farming, a significant source of methane and land use emissions (Fig. 5.2; Shah and Krishnakumar, 2025). Applications of nanotechnology in food systems must be supported by careful consideration of safety, regulatory frameworks, and public acceptance. Comprehensive toxicological evaluations and the development of standardized regulations are crucial to ensure that NMs do not pose risks to human health or the environment. Adopting sustainable synthesis approaches and conducting life cycle assessments of NMs are essential to confirm their overall benefits for climate change mitigation (Vaidya *et al.*, 2024; Shah and Krishnakumar, 2025).

### 5.2.1 Enhanced crop productivity and resource efficiency

Enhanced crop productivity and resource efficiency are critical goals in modern agriculture, driven by the need to meet the growing global food demand while minimizing environmental impact. Improving crop yields involves optimizing inputs such as water, nutrients, and energy, alongside adopting sustainable practices that conserve resources and maintain soil health (Diallo and Brinker, 2011; Li *et al.*, 2024). Agriculture technologies, including GPS-guided machinery, remote sensing, and data analytics, enable farmers to apply water and fertilizers more efficiently, reducing waste and runoff. This

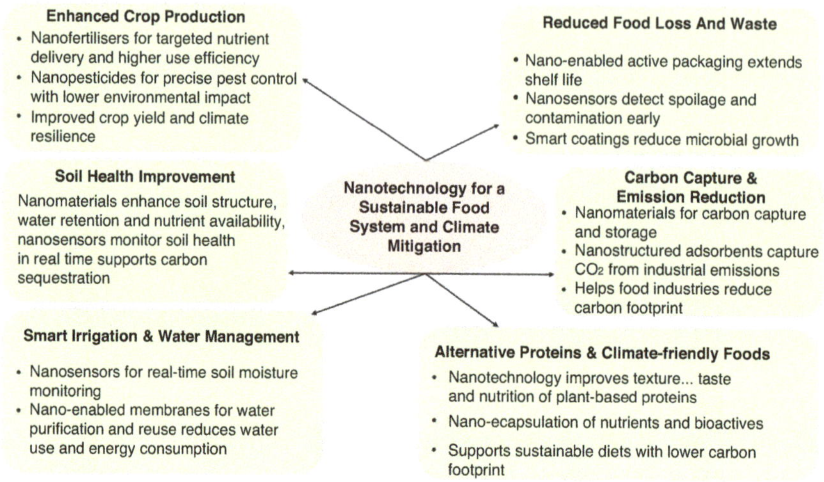


**Fig. 5.2.** Applications of nanotechnology in food systems for climate mitigation.

targeted input management enhances nutrient uptake in plants, resulting in higher biomass and grain production (Huang and Wang, 2024). Integrating crop rotation, cover cropping, and conservation tillage helps preserve soil structure and fertility, improving water retention and reducing erosion. Water-use efficiency is enhanced through advanced irrigation techniques such as drip irrigation and sprinkler systems, which deliver water directly to the root zone and reduce evaporation losses (Godfray *et al.*, 2024; Hiywotu, 2025). Crop varieties bred for drought tolerance and nutrient-use efficiency further support productivity under limited resource conditions. Moreover, organic amendments and biofertilizers can improve nutrient cycling and soil microbial activity, contributing to sustainable crop growth (Godfray *et al.*, 2024; Rani *et al.*, 2025a). Resource efficiency is also linked to energy use; adopting renewable energy sources and energy-efficient machinery reduces the carbon footprint of agricultural operations. Digital tools and automation optimize planting, harvesting, and postharvest handling, minimizing input costs and losses (Huang and Wang, 2024; Hiywotu *et al.*, 2025).

### 5.2.2 Reduction of postharvest losses

Postharvest losses in the food supply chain contribute significantly to global food insecurity and increased GHG emissions due to wasted resources and decomposing biomass. Nanotechnology offers innovative solutions to reduce these losses by improving food preservation, extending shelf life, and enhancing food safety, thereby supporting climate mitigation efforts (Chamoli *et al.*, 2025). Nanomaterials can be integrated into packaging to provide antimicrobial properties, oxygen scavenging, and controlled release of preservatives, reducing spoilage and contamination (Li *et al.*, 2024). Nano-enabled sensors embedded in packaging or storage facilities allow real-time food quality monitoring, detecting spoilage or contamination early, and reducing waste. Furthermore, nanoscale coatings can be applied to fruits and vegetables to retard respiration and moisture loss, effectively prolonging freshness. Using nano-emulsions improves the delivery and efficacy of natural preservatives and antioxidants, reducing the need for synthetic chemicals. Nanotechnology facilitates improved cold chain logistics through nanosensors that monitor temperature and humidity, minimizing losses during transport. Reducing spoilage through nanotechnology indirectly lowers methane emissions generated from the decomposition of organic waste in landfills (Huang and Wang, 2024; Hiywotu *et al.*, 2025).

### 5.2.3 Development of sustainable food packaging

The development of sustainable food packaging has emerged as a critical focus area within the broader context of nanotechnology applications in food systems, particularly aimed at climate change mitigation (Ameta *et al.*, 2024). Traditional food packaging materials (Rani *et al.*, 2025), predominantly plastics derived from fossil fuels, contribute significantly to environmental pollution and GHG emissions. Nanotechnology offers innovative solutions to enhance the sustainability and functionality of food packaging by improving material properties, reducing waste, and extending shelf life, collectively reducing food supply chains' carbon footprint. Nanomaterials such as nanoclays, nanocellulose, Ag NPs, and nanosilica are increasingly incorporated into biodegradable and bio-based polymers to create advanced packaging materials (Table 5.2; Ameta *et al.*, 2024). These nanocomposites exhibit superior mechanical strength, barrier properties against oxygen, moisture, UV light, and antimicrobial activity. Improving the protective qualities of packaging minimizes food spoilage and waste, leading to less resource consumption and lower emissions related to food production and disposal (George *et al.*, 2022; Ameta *et al.*, 2024; Chamoli *et al.*, 2025). Moreover, nanotechnology enables the development of active and intelligent packaging systems. Active packaging, such as oxygen scavengers or antimicrobial agents, can release or absorb substances to maintain food freshness. Smart packaging systems, often integrated with nanosensors, monitor food quality by detecting spoilage indicators or contamination in real time, thereby reducing unnecessary disposal and enabling informed consumption decisions (Kang *et al.*, 2017; George *et al.*, 2022; Chamoli *et al.*, 2025).

### 5.2.4 Improved food-processing efficiency

Nanotechnology has emerged as a transformative tool in food processing, offering significant efficiency, safety, and sustainability improvements (Chang *et al.*, 2024). Through nanoscale engineering of materials, food systems can be designed to optimize processing techniques, minimize waste, and mitigate climate impacts. The integration of nanotechnology into food processing improves key stages, including preservation, packaging, quality control, and nutrient delivery, thereby contributing to reduced GHG emissions and more efficient resource utilization (Gu *et al.*, 2017; Tehri *et al.*, 2025). One key area where nanotechnology improves processing efficiency is the development of

**Table 5.2.** Nanomaterials used in sustainable food packaging, their functions, benefits for climate mitigation, and applications.

| Nanomaterial (typical host polymer/coating) | Primary functions in packaging | Climate-mitigation benefit (mechanism) | Application examples | References |
|---|---|---|---|---|
| Cellulose Nanofibrils (CNF) (CNF/CNC) in biopolymers (PLA, starch, Polyhydroxyalkanoates (PHA)) and paper coatings | Improves $O_2$/$CO_2$ and water-vapor barrier; mechanical reinforcement; supports active agents' antioxidants/antimicrobials; can enable smart/indicator films | Extends shelf life → less food waste; enables thinner films (material reduction); supports bio-based/compostable substrates | CNF-reinforced PLA films; nanocellulose coatings on paper to replace plastic liners; intelligent freshness indicators | Rani *et al.*, 2025 |
| Nanoclays (e.g. montmorillonite, halloysite) in PLA, Poly(butylene adipate-co-terephthalate) (PBAT), starch, Low-density polyethylene (LDPE) | Strong barrier via tortuous diffusion path; thermal/mechanical enhancement; carrier for active agents (Essential oils (Eos), antimicrobials) | Shelf-life extension → less waste; downgauging; potential to improve biodegradation control of bioplastics | PLA/Montmorillonite (MMT) or PBAT/MMT nanocomposites for fresh produce; LDPE-nanoclay films with reduced $O_2$ permeability | Bumbudsanpharoke *et al.*, 2017 |
| ZnO nanoparticles in biopolymers (PLA, chitosan) and PE | Broad-spectrum antimicrobial; UV blocking; barrier/mechanical gains at low loadings | Reduces microbial spoilage → less waste; reduces need for refrigeration/UV degradation of foods; thinner films possible | Chitosan/ZnO coatings for fruits/meat; PLA/ZnO films for UV-sensitive foods | Vagena *et al.*, 2024 |
| $TiO_2$ nanoparticles (often photoactive) in polyolefins/biopolymers | Photocatalytic antimicrobial under UV/visible activation; UV barrier; surface self-cleaning | Limits surface contamination → waste reduction; potential to allow milder sanitation regimes | UV-activated $TiO_2$ films for bakery/produce; $TiO_2$-coated trays | Prakash *et al.*, 2022 |

| | | | | |
|---|---|---|---|---|
| Copper oxide (CuO) or copper nanoparticles in biopolymer matrices | Antimicrobial; antioxidant; mechanical reinforcement | Shelf-life extension and reduced recalls; potential downgauging | CuO-reinforced starch/PLA films used to protect juices/ready-to-drink items from microbial spoilage | Ameta *et al.*, 2024 |
| Graphene oxide (GO)/reduced GO in biopolymers | Exceptional gas barrier; UV blocking; mechanical stiffness; possible antimicrobial surface effects | Enables ultra-thin barrier layers → less packaging mass; extends shelf life of high respiration produce | GO/biopolymer coatings for meat/fish MAP films; GO/starch high-barrier pouches | Perera *et al.*, 2023; Chandra *et al.*, 2025 |
| Silica ($SiO_2$) nanoparticles and hybrid silica layers | Gas barrier; scratch/abrasion resistance; anti-fog | Downgauging via thin barrier coats maintains visibility/quality → reduced discard | $SiO_2$-hybrid coatings on Polyethylene terephthalate (PET)/PLA lids and films | Ramezanzadeh *et al.*, 2016 |
| Starch nanocrystals/nanostarch in starch/PLA matrices | Reinforcement; improved barrier; carrier for bioactives | Biobased content increase; shelf-life gains; potential compostability | Nanostarch-reinforced PLA trays and biodegradable pouches | Marta *et al.*, 2022; Hussain *et al.*, 2024 |
| Chitosan nanoparticles (often loaded with essential oils or bacteriocins) | Intrinsic antimicrobial/antioxidant; film-forming; carrier for actives with controlled release | Shelf-life extension; can replace synthetic preservatives; reduces product loss | Chitosan-NP coatings on fruits, cheese; EO-loaded chitosan films for meat | Khorshidi *et al.*, 2021 |
| Essential oil nano-emulsions (as active nanofillers or coatings) | Antimicrobial/antioxidant release; odor-control | Cuts fungal/bacterial spoilage; can enable fewer cold chain demands | Cinnamon/oregano-oil nano-emulsions in biopolymer films for fresh produce/meat | Maurya *et al.*, 2021; Hanan *et al.*, 2024 |

*Continued*

**Table 5.2.** Continued

| Nanomaterial (typical host polymer/coating) | Primary functions in packaging | Climate-mitigation benefit (mechanism) | Application examples | References |
|---|---|---|---|---|
| PBAT/PLA bio-nanocomposites (multi-filler systems incl. nanoclay, ZnO, $SiO_2$) | Balanced barrier + toughness; active/UV functions | Compostable options with performance → material reduction + waste cuts | PBAT nanocomposites for bakery/snack films; PLA multilayers | Venkatesan and Rajeswari, 2017; Jia *et al.*, 2024 |
| Cellulose–clay hybrid nanocomposites | Synergistic barrier (cellulose + clay); mechanical reinforcement | Enables paper/biopolymer replacement of multilayer plastics; shelf-life extension | Paperboard with Nanocomposite (NC)/MMT coat replacing plastic liners | Mirmehdi *et al.*, 2018; Adak *et al.*, 2025 |
| General nanocomposite approach (overview/reference category) | Low loading nanofillers upgrade barrier, mechanical, thermal, and can embed actives | Cross-cutting route to thinner films and longer shelf life across formats | Multiple polymer families for MAP lids, pouches, trays | Sarfraz *et al.*, 2020; Moradi *et al.*, 2023 |
| Sustainability/roadmap overviews (context) | State-of-the-art on sustainable nanopackaging and safety considerations | Aligns material choices to waste reduction and circularity | Strategy papers guiding selection and LCA (Life Cycle Assessment) | Chang *et al.*, 2024 |

nanosensors. These sensors enable real-time food quality and safety monitoring by detecting pathogens, contaminants, or spoilage indicators with high sensitivity. This technology reduces the need for excessive preservatives and lowers food wastage, indirectly decreasing emissions from food production and disposal (Sun *et al.*, 2023; Chang *et al.*, 2024).

Nano-encapsulation techniques allow for the precise delivery of nutrients, flavors, and preservatives, improving food quality and shelf life while minimizing the quantity of additives required (Chamoli *et al.*, 2025). This targeted delivery system reduces energy consumption during processing and transportation and curbs nutrient loss, contributing to a more sustainable food supply chain (Kingra and Misra, 2022; Tehri *et al.*, 2025). Nanomaterials also enhance food packaging by creating barriers that improve mechanical strength, thermal resistance, and antimicrobial properties. Nanocomposite films reduce the permeability of gases such as oxygen and carbon dioxide, extending product shelf life and reducing spoilage (Chang *et al.*, 2024; Yadav *et al.*, 2025). Preventing premature spoilage, these materials contribute to reducing food waste and its associated carbon footprint (Tehri *et al.*, 2025). In terms of processing equipment, nanocoatings on surfaces reduce fouling and microbial contamination, improving hygiene and lowering cleaning requirements (Vieira *et al.*, 2022; Panda *et al.*, 2024). This leads to energy savings and reduces water use, contributing to more sustainable operations. Nanotechnology can improve heat and mass transfer processes, optimizing cooking and drying operations to lower energy consumption (Gu *et al.*, 2017; Chang *et al.*, 2024).

### 5.2.5 Development of functional foods and nutraceuticals

Nanotechnology has emerged as a transformative approach in the development of functional foods and nutraceuticals, offering promising solutions to enhance food quality, safety, and environmental sustainability. Functional foods are designed to provide health benefits beyond basic nutrition, while nutraceuticals include products derived from food sources with extra health-promoting properties (Vieira *et al.*, 2022; Aguilar-Pérez *et al.*, 2023). The integration of nanotechnology in these domains supports climate mitigation by improving resource efficiency, reducing waste, and enhancing the bioavailability of beneficial compounds, thus contributing to sustainable food systems (Singh *et al.*, 2019; Abdulqahar and Hussein, 2023). Nanotechnology enables the encapsulation, protection, and controlled release of bioactive compounds such as vitamins, antioxidants, polyphenols, and omega-3 fatty acids. This improves their stability during processing and storage and enhances their absorption in the human body (Heydari, 2024). For example, nano-emulsions and liposomes can be used to deliver hydrophobic nutrients effectively, leading to better health outcomes and reducing the need for excessive supplementation (Sabliov *et al.*, 2015; Vieira *et al.*, 2022). This precision in delivery reduces the environmental burden associated with overproduction and wastage of nutrients. Moreover, nano-encapsulation techniques facilitate the use of agro-food by-products, converting waste into value-added ingredients rich in bioactive compounds (Abdulqahar and Hussein,

2023). This supports circular economy principles and decreases GHG emissions associated with waste disposal. The use of nanosensors and smart packaging integrated with NMs allows real-time monitoring of food quality and shelf life, reducing spoilage and food loss, which are significant contributors to global carbon emissions (Sabliov *et al.*, 2015; Tehri *et al.*, 2025).

## 5.3 Challenges and Future Perspectives

Nanotechnology presents significant potential to address the climate impact of food systems by enhancing efficiency, reducing waste, and minimizing GHG emissions. The integration of nanotechnology in agriculture, food processing, packaging, and distribution is accompanied by a series of challenges that must be addressed to harness its benefits fully for sustainable food systems. One of the primary challenges is the safety and regulatory uncertainty surrounding using NMs in food systems (Tehri *et al.*, 2025). Owing to their small size and high reactivity, Nanoparticles may pose unknown toxicological risks to humans and the environment. Current regulations are insufficient or inconsistent across countries, leading to safety assessment and risk management gaps. Standardized protocols for evaluating nanomaterial toxicity, bioaccumulation, and environmental fate are crucial for safe implementation (Sabliov *et al.*, 2015; Tehri *et al.*, 2025).

Another challenge lies in the scalability and economic feasibility of nanotechnology applications. Although laboratory and pilot-scale studies have demonstrated the benefits of nanotech-based innovations such as nanofertilizers, nanopesticides, and nano-encapsulated nutrients, their commercial-scale cost-effectiveness remains uncertain. The manufacturing processes for NMs are often complex and energy-intensive, which could offset some environmental benefits if not optimized. The high cost may limit adoption by smallholder farmers and producers in developing countries, who constitute a large portion of the global food supply chain (Mustapha *et al.*, 2025). Public perception and acceptance of nanotechnology in food are another significant hurdle. Consumers tend to be wary of "nano" products due to fears of health risks and ethical concerns. Transparency in labeling, consumer education, and engagement in dialogues about the benefits and risks of nanotechnology are essential to build trust and market acceptance. Research gaps also exist in understanding the long-term environmental impacts of NMs used in agriculture and food packaging. The persistence, transformation, and bioaccumulation of NMs in soil and water ecosystems are poorly understood. There is a need for interdisciplinary research combining nanotechnology, environmental science, and toxicology to evaluate the life cycle impacts and develop eco-friendly nano-enabled solutions (Tehri *et al.*, 2025).

Despite these challenges, future perspectives for nanotechnology in mitigating the climate impact of food systems are promising (Fig. 5.3). Advances in green synthesis methods aim to produce NMs using sustainable and low-energy processes, reducing environmental footprints. Smart nanosensors can enable precision agriculture by real-time monitoring soil nutrients, pest infestations, and plant health, optimizing input use and reducing emissions (Mustapha *et al.*, 2025). Nano-enabled food packaging that improves shelf

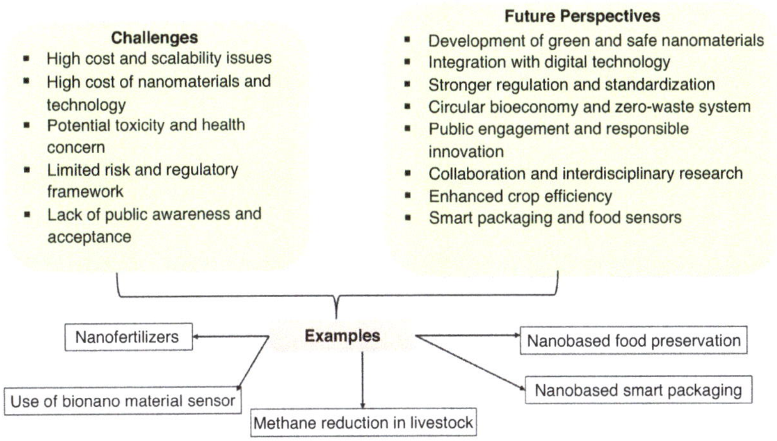


**Fig. 5.3.** Nanotechnology in addressing the climate impact of food systems and challenges, and future perspectives.

life and reduces food spoilage can significantly decrease food waste, a major contributor to global GHG emissions. Nanotechnology can facilitate the design of alternative protein sources and functional foods with lower environmental impacts, supporting dietary shifts needed for climate mitigation.

## 5.4 Conclusion

Nanotechnology holds significant promise for reducing the climate impact of food systems by improving sustainability, lowering GHG emissions and enhancing resource efficiency across the supply chain. From agricultural production to waste management, food systems are major contributors to climate change. Nanotechnology-based innovations offer targeted solutions to mitigate these impacts. In agriculture, nanofertilizers and nanopesticides enable controlled release and precise delivery of nutrients and chemicals, reducing excessive application, nutrient runoff, and nitrous oxide emissions. Nanosensors further support precision farming by providing real-time monitoring of soil conditions, water quality, and crop health, thereby optimizing input use and minimizing waste. In food processing and preservation, NMs enhance energy efficiency and extend shelf life, reducing food loss and associated emissions. Nano-enabled packaging with antimicrobial and barrier properties improves food safety and reduces dependence on refrigeration and transportation, lowering the carbon footprint. Nanocatalysts and nanostructured membranes facilitate food waste valorization, converting waste into bioenergy and valuable by-products, thus supporting circular economy approaches. The widespread adoption of nanotechnology requires careful risk assessment, regulatory oversight, and life cycle analysis to ensure environmental and human safety. The development of biodegradable NMs and green synthesis methods is essential. Nanotechnology offers a comprehensive pathway toward more sustainable and climate-resilient food systems.

# References

Abdel-Karim, R. (2024) Nanotechnology-enabled biosensors: A review of fundamentals, materials, applications, challenges, and future scope. *Biomedical Materials & Devices* 2(2), 759–777. DOI: 10.1007/s44174-023-00147-z.

Abdulqahar, F.W. and Hussein, F.F. (2023) Utilization of nano-form functional foods and nutraceuticals and their significance in health boosting and the COVID-19's challenge. In: *IOP Conference Series: Earth and Environmental Science*, IOP Publishing, p. 012172 (Vol. 1252, no. 1). DOI: 10.1088/1755-1315/1252/1/012172.

Abouzeinab, N.S., Kahil, N., Fakhruddin, N., Awad, R. and Khalil, M.I. (2023) Intraperitoneal hepato-renal toxicity of zinc oxide and nickel oxide nanoparticles in male rats: Biochemical, hematological, and histopathological studies. *EXCLI Journal* 22, 619. Available at: https://pmc.ncbi.nlm.nih.gov/articles/PMC10471841/

Adak, B., Baidya, S. and Teramoto, Y. (2025) Biopolymers and their nanocomposites coated paper-based high barrier and sustainable food packaging materials. *Carbohydrate Polymers* 123966.

Adeyeye, S.A.O., Babu, A.S. and Ganesh, P.S. (2023) Starch nanocrystal and its food packaging applications. *Current Research in Nutrition and Food Science Journal* 11(1), 01–21. DOI: 10.12944/CRNFSJ.11.1.1.

Aggarwal, S., Sehgal, S. and Tandon, W. (2021) Biosensors: Types and applications in the food processing industry. *Journal of Postharvest Technology* 9(1), 1–19. pht.in/MenuscriptFile/7d7d02d5-3d7b-43ae-af20-53bacfcdf48f.pdf

Agnihotri, S. (2025) Nano-emulsions as advanced delivery systems of bioactive compounds for sustainable food preservation applications. *Biocatalysis and Agricultural Biotechnology* 103702. DOI: 10.1016/j.bcab.2025.103702.

Aguilar-Pérez, K.M., Ruiz-Pulido, G., Medina, D.I., Parra-Saldivar, R. and Iqbal, H.M. (2023) Insight of nanotechnological processing for nano-fortified functional foods and nutraceutical—opportunities, challenges, and future scope in food for better health. *Critical Reviews in Food Science and Nutrition* 63(20), 4618–4635. DOI: 10.1080/10408398.2021.2004994.

Ahari, H. and Soufiani, S.P. (2021) Smart and active food packaging: Insights in novel food packaging. *Frontiers in Microbiology* 12, 657233. DOI: 10.3389/fmicb.2021.657233.

Ahmad, S.S., Yousuf, O., Islam, R.U. and Younis, K. (2021) Silver nanoparticles as an active packaging ingredient and its toxicity. *Packaging Technology and Science* 34(11–12), 653–663. DOI: 10.1002/pts.2603.

Ahmad, M.A., Adeel, M., Shakoor, N., Javed, R., Ishfaq, M. *et al.* (2023) Modifying engineered nanomaterials to produce next-generation agents for environmental remediation. *Science of the Total Environment* 894, 164861. DOI: 10.1016/j.scitotenv.2023.164861.

Ahmad, K., Li, Y., Tu, C., Guo, Y., Yang, X. *et al.* (2024) Nanotechnology in food packaging with implications for sustainable outlook and safety concerns. *Food Bioscience* 58, 103625. DOI: 10.1016/j.fbio.2024.103625.

Ahmad, K., Ahmad, R., Faizan, M., Ali, F., Yousaf, M.M. *et al.* (2025) A review of nanotechnology in food industry with implications for viable outlook and safety issues. *Hybrid Advances* 100487. DOI: 10.1016/j.hybadv.2025.100487.

Ahmadi, N., Ahari, H., Anvar, A., Khosravi-Darani, K. and Gharachorloo, M. (2024) Polycaprolactone (PCL)-based films integrated with hairy cellulose nanocrystals and silver nanoparticles for active tilapia packaging applications. *Food Chemistry: X* 22, 101490. DOI: 10.1016/j.fochx.2024.101490.

Ahmed, S.R., Kim, J., Suzuki, T., Lee, J. and Park, E.Y. (2022) Nanomaterial-based biosensors for detection of foodborne pathogens. *Trends in Biotechnology* 40(3), 341–357. DOI: 10.1111/1541-4337.12576.

Akanda, M., Mithu, M.S.H. and Douroumis, D. (2023) Solid lipid nanoparticles: An effective lipid-based technology for cancer treatment. *Journal of Drug Delivery Science and Technology* 86, 104709. DOI: 10.1016/j.jddst.2023.104709.

Alberti, G., Zanoni, C., Magnaghi, L.R. and Biesuz, R. (2021) Gold and silver nanoparticle-based colorimetric sensors: New trends and applications. *Chemosensors* 9(11), 305. DOI: 10.3390/chemosensors9110305.

Al-Doaiss, A.A., Jarrar, Q., Alshehri, M. and Jarrar, B. (2020) *In vivo* study of silver nanomaterials' toxicity with respect to size. *Toxicology and Industrial Health* 36(8), 540–557. DOI: 10.1177/0748233720937201.

Alfei, S. (2020) Nanotechnology applications to improve the solubility of bioactive constituents of foods for health-promoting purposes. In: *Nano-Food Engineering: Volume One*. Springer International Publishing, Cham, pp. 189–257. DOI: 10.1007/978-3-030-44552-2_8.

Ali, S., Chen, X., Ahmad, S., Shah, W., Shafique, M. *et al.* (2023) Advancements and challenges in phytochemical-mediated silver nanoparticles for food packaging: Recent review (2021–2023). *Trends in Food Science & Technology* 141, 104197. DOI: 10.1016/j.tifs.2023.104197.

Aliabbasi, N. and Emam-Djomeh, Z. (2024) Application of nanotechnology in dairy desserts and ice cream formulation with the emphasis on textural, rheological, antimicrobial, and sensory properties. *eFood* 5(4), e170. DOI: 10.1002/efd2.170.

Alquraishi, M.I., Alfadda, N.A., Alabdullatif, W.A., Veeramani, C., El Newehy, A.S. *et al.* (2025) Fruit and vegetable preservation from classical to advanced nanotechnology: An overview of efficacy and health concerns. *Food Chemistry: X* 30, 102984. DOI: 10.1016/j.fochx.2025.102984.

Alshammari, A. (2023) Development of a ZnO-based aptamer system for biosensing applications. Doctoral dissertation, University of Liverpool. Available at: https://livrepository.liverpool.ac.uk/3178616/1/201374259_Jan2024.pdf

Altemimi, A.B., Farag, H.A.M., Salih, T.H., Awlqadr, F.H., Al-Manhel, A.J.A. *et al.* (2024) Application of nanoparticles in human nutrition: A review. *Nutrients* 16(5), 636.

Ameta, S.K., Rai, A.K., Hiran, D., Ameta, R. and Ameta, S.C. (2020) Use of nanomaterials in food science. In: *Biogenic Nano-Particles and Their Use in Agro-Ecosystems*. Springer Singapore, Singapore, pp. 457–488. DOI: 10.1007/978-981-15-2985-6_24.

Ameta, R.K., Malik, P., Korgaokar, S., Vanzara, P. and Soni, K. (2024) Contemporary advances in the plant resources mediated synthesis of copper oxide nanoparticles: Insights on structure-function-workability understanding. *Plant Nano Biology* 7, 100065. DOI: 10.1016/j.plana.2024.100065.

Amir, M., Singh, P. and Bano, N. (2025) Food waste and its biotechnological application. In: *Resource Recycling and Management of Food Waste*. Springer, Cham, pp. 305–318. DOI: 10.1007/978-3-031-86688-3_13.

Angelopoulou, P., Giaouris, E. and Gardikis, K. (2022) Applications and prospects of nanotechnology in food and cosmetics preservation. *Nanomaterials* 12(7), 1196.

Ankiel, M., Halagarda, M., Piekara, A., Sady, S., Żmijowska, P. *et al.* (2025) Role of certifications and labelling in ensuring authenticity and sustainability of fermented milk products. *Sustainability* 17(18), 8398. DOI: 10.3390/su17188398.

Antiochia, R., Vinci, G. and Gorton, L. (2013) Rapid and direct determination of fructose in food: A new osmium-polymer mediated biosensor. *Food Chemistry* 140(4), 742–747. DOI: 10.1016/j.foodchem.2012.11.023.

Apoorva, M.S., Kumar, S. and Bhatia, A. (2023) Nanotechnological interventions for climate change mitigation: A. *Journal of Agricultural Physics* 23(1), 1–14.

Ardhi, A., Raharjo, S., Sudjarwo, W.A.A. and Schreiner, M. (2025) Nanostructured lipid carriers for enhanced bioaccessibility of thymoquinone and its application in pure and sparkling apple juice fortification. *LWT* 221, 117578. DOI: 10.1016/j.lwt.2025.117578.

Ashfaq, A., Khursheed, N., Fatima, S., Anjum, Z. and Younis, K. (2022) Application of nanotechnology in food packaging: Pros and cons. *Journal of Agriculture and Food Research* 7, 100270.

Ashfaq, R., Rasul, A., Asghar, S., Kovács, A., Berkó, S. *et al.* (2023) Lipid nanoparticles: An effective tool to improve the bioavailability of nutraceuticals. *International Journal of Molecular Sciences* 24(21), 15764. DOI: 10.3390/ijms242115764.

Ashraf, S. A. (2025). Food fortification as a sustainable global strategy to mitigate micronutrient deficiencies and improve public health. *Discover Food* 5(1), 201.

Atilgan, H., Unal, B., Yalcinkaya, E.E., Evren, G., Atik, G. *et al.* (2023) Development of an enzymatic biosensor using glutamate oxidase on organic–inorganic-structured, electrospun nanofiber-modified electrodes for monosodium glutamate detection. *Biosensors* 13(4), 430. DOI: 10.3390/bios13040430.

Auclair, J. and Gagné, F. (2022) Shape-dependent toxicity of silver nanoparticles on freshwater cnidarians. *Nanomaterials* 12(18), 3107. DOI: 10.3390/nano12183107.

Awashra, M. and Młynarz, P. (2023) The toxicity of nanoparticles and their interaction with cells: An *in vitro* metabolomic perspective. *Nanoscale Advances* 5(10), 2674–2723. DOI: 10.1039/D2NA00534D.

Awasthi, S., Srivastava, A., Kumar, D., Pandey, S.K., Mubarak, N.M. *et al.* (2024) An insight into the toxicological impacts of carbon nanotubes (CNTs) on human health: A review. *Environmental Advances* 18, 100601. DOI: 10.1016/j.envadv.2024.100601.

Awlqadr, F.H., Altemimi, A.B., Qadir, S.A., Salih, T.A.H., Alkanan, Z.T. *et al.* (2024) Emerging trends in nano-sensors: A new frontier in food safety and quality assurance. *Heliyon*. DOI: 10.1016/j.heliyon.2024.e41181.

Awlqadr, F.H., Altemimi, A.B., Qadir, S.A., Salih, T.A.H., Alkanan, Z.T. *et al.* (2025) Emerging trends in nano-sensors: A new frontier in food safety and quality assurance. *Heliyon* 11(1). DOI: 10.1016/j.heliyon.2024.e41181.

Azeredo, H.M., Rosa, M.F. and Mattoso, L.H.C. (2017) Nanocellulose in bio-based food packaging applications. *Industrial Crops and Products* 97, 664–671. DOI: 10.1016/j.indcrop.2016.03.013.

Bandyopadhyay, J. and Ray, S.S. (2023) Nanotechnology in innovative food preservation and packaging. In: *Emerging Technologies in Food Preservation*. CRC Press, pp. 349–376. DOI: 10.1201/9781003147978-13.

Baumgartner, J., Winkler, H.C., Zandberg, L., Tuntipopipat, S., Mankong, P. *et al.* (2022) Iron from nanostructured ferric phosphate: Absorption and biodistribution in mice and bioavailability in iron deficient anemic women. *Scientific Reports* 12(1), 2792. DOI: 10.1038/s41598-022-06701-x.

Berthing, T., Lard, M., Danielsen, P.H., Abariute, L., Barfod, K.K. *et al.* (2023) Pulmonary toxicity and translocation of gallium phosphide nanowires to secondary organs following pulmonary exposure in mice. *Journal of Nanobiotechnology* 21(1), 322. DOI: 10.1186/s12951-023-02049-0.

Bhardwaj, N., Gianni, E., Bhardwaj, S., Gulati, A. and Gulati, S. (2022) Development of microbial-based nano-sensors and their applications: The state of the art. In: Kumar, V. and Singh, M. (eds) *Microbial Nanotechnology: Advances in Agriculture, Industry and Health Sectors*. De Gruyter, Berlin, Boston, pp. 211–242.

Biswas, R., Alam, M., Sarkar, A., Haque, M.I., Hasan, M.M. *et al.* (2022) Application of nanotechnology in food: Processing, preservation, packaging, and safety assessment. *Heliyon* 8, 11. DOI: 10.1016/j.heliyon.2022.e11795.

Blaznik, U., Krušič, S., Hribar, M., Kušar, A., Žmitek, K. *et al.* (2021) Use of food additive titanium dioxide (E171) before the introduction of regulatory restrictions due to concern for genotoxicity. *Foods* 10(8), 1910. DOI: 10.3390/foods10081910.

Boro, P.R., Borthakur, P.P. and Baruah, E. (2025) Advancements in optical biosensor technology for food safety and quality assurance. *Engineering Proceedings* 106(1), 6.

Boshra, M.H., El-Housseiny, G.S., Farag, M.M. and Aboshanab, K.M. (2024) Innovative approaches for mycotoxin detection in various food categories. *AMB Express* 14(1), 7.

Bouwmeester, H., Dekkers, S., Noordam, M.Y., Hagens, W.I., Bulder, A.S. *et al.* (2009) Review of health safety aspects of nanotechnologies in food production. *Regulatory Toxicology and Pharmacology* 53(1), 52–62. DOI: 10.1016/j.yrtph.2008.10.008.

Boz, Z., Welt, B.A., Brecht, J.K., Pelletier, W., McLamore, E. *et al.* (2018) Review of challenges and advances in modification of food package headspace gases. *Journal of Applied Packaging Research* 10(1), 5. Available at: https://repository.rit.edu/japr/vol10/iss1/5/

Brito, C.L., Silva, J.V., Gonzaga, R.V., La-Scalea, M.A., Giarolla, J. *et al.* (2024) A review on carbon nanotubes family of nanomaterials and their health field. *ACS Omega* 9(8), 8687–8708.

Brody, A.L., Bugusu, B., Han, J.H., Sand, C.K. and McHugh, T.H. (2008) Innovative food packaging solutions. *Journal of Food Science* 73(8), 107–116. Available at: https://www.technologyed.com/oldfiles/courses/c230/1.pdf

Bryksa, B.C. and Yada, R.Y. (2012) Challenges in food nanoscale science and technology. *Journal of Food and Drug Analysis* 20(1), 38. DOI: 10.38212/2224-6614.2104.

Bumbudsanpharoke, N., Lee, W., Choi, J.C., Park, S.J., Kim, M. *et al.* (2017) Influence of montmorillonite nanoclay content on the optical, thermal, mechanical, and barrier properties of low-density polyethylene. *Clays and Clay Minerals* 65(6), 387–397.

Camargo, J.R., Orzari, L.O., de Souza Rodrigues, J., Lima, L.F., Paixão, T.R.L.C. *et al.* (2024) Advancements in disposable electrochemical systems for sustainable agriculture monitoring: Trends, gaps, and applied examples. *TrAC Trends in Analytical Chemistry* 117968. DOI: 10.1016/j.trac.2024.117968.

Carbone, M., Donia, D.T., Sabbatella, G. and Antiochia, R. (2016) Silver nanoparticles in polymeric matrices for fresh food packaging. *Journal of King Saud University-Science* 28(4), 273–279. DOI: 10.1016/j.jksus.2016.05.004.

Chamoli, S., Shimali, 0., Chamoli, A., Karki, K., Kumar, R. *et al.* (2025) Recent advances in sustainable strategies for the development of innovative nanobiocatalysts using immobilized β-glucosidase for industrial applications. *Critical Reviews in Biotechnology* 1–20. DOI: 10.1080/07388551.2025.2517714.

Chandra, D.K., Kumar, A. and Mahapatra, C. (2025) Advanced nano-enhanced bioplastics for smart food packaging: Enhancing functionalities and sustainability. *Cleaner and Circular Bioeconomy* 100140.

Chang, J.H., Kumar, M., Selvaraj, S., Samuel, M.S., Ethiraj, S. *et al.* (2024) Green energy breakthroughs: Harnessing nano-catalysts and enzymatic catalysts for bioenergy generation. *Industrial Crops and Products* 215, 118527. DOI: 10.1016/j.indcrop.2024.118527.

Chaudhary, P. and Janmeda, P. (2025) Industrial application of bio-nanomaterial in food industry. In: *Bio-Nanomaterials in Environmental Remediation: Industrial Applications*. pp. 191–224. DOI: 10.1002/9783527848546.ch8.

Chaudhary, Q., Scotter, M., Blackburn, J., Ross, B., Boxall, A. *et al.* (2008) Applications and implications of nanotechnologies for the food sector. *Food Additives and Contaminants* 25(3), 241–258. DOI: 10.1080/02652030701744538.

Chaudhary, R., Gaur, N. and Diwan, B. (2022) Nano-encapsulation for production of fermented foods and pigments. In: *Smart Nanomaterials for Bioencapsulation*. Elsevier, pp. 237–251. DOI: 10.1016/B978-0-323-91229-7.00013-1.

Chauhan, C., Dhir, A., Akram, M.U. and Salo, J. (2021) Food loss and waste in food supply chains. A systematic literature review and framework development approach. *Journal of Cleaner Production* 295, 126438. DOI: 10.1016/j.jclepro.2021.126438.

Chausali, N., Saxena, J. and Prasad, R. (2023) Nanotechnology as a sustainable approach for combating the environmental effects of climate change. *Journal of Agriculture and Food Research* 12, 100541. DOI: 10.1016/j.jafr.2023.100541.

Chen, Q., Riviere, J.E. and Lin, Z. (2022a) Toxicokinetics, dose–response, and risk assessment of nanomaterials: Methodology, challenges, and future perspectives. *Wiley Interdisciplinary Reviews: Nanomedicine and Nanobiotechnology* 14(6), e1808. DOI: 10.1002/wnan.1808.

Chen, S., Hwang, M.T., Wang, J., Ganguli, A., Park, I. *et al.* (2022b) December, Advances in electronic nano-biosensors and new frontiers in bioengineering. In: *2022 International Electron Devices Meeting (IEDM)*, IEEE, pp. 17–1. DOI: 10.1109/IEDM45625.2022.10019512.

Chen, X., Wang, Y. and Liu, H. (2023a) AI-driven prediction models for food spoilage and shelf-life optimization. *Journal of Food Engineering* 345, 111234.

Chen, B.Y., Hong, S.Y., Wang, H.M., Shi, Y., Wang, P. *et al.* (2023b) The subacute toxicity and underlying mechanisms of biomimetic mesoporous polydopamine nanoparticles. *Particle and Fibre Toxicology* 20(1), 38. DOI: 10.1186/s12989-023-00548-4.

Chen, Y., Wang, Y., Zhang, Y., Wang, X., Zhang, C. *et al.* (2024) Intelligent biosensors promise smarter solutions in food safety 4.0. *Foods* 13(2), 235. DOI: 10.3390/foods13020235.

Cheng, H., Chen, L., McClements, D.J., Xu, H., Long, J. *et al.* (2024) Recent advances in the application of nanotechnology to create antioxidant active food packaging materials. *Critical Reviews in Food Science and Nutrition* 64(10), 2890–2905. DOI: 10.1080/10408398.2022.2128035.

Cheng, J., Li, X., Ke, T., Gong, J., Jiang, T. *et al.* (2025) Advances in polysaccharide-based probiotic delivery systems: A review. *Carbohydrate Polymer Technologies and Applications* 100804. DOI: 10.1016/j.carpta.2025.100804.

Chhabra, N., Shiriskar, J. and Srinivasan, G. (2025) Current and future market of the dietary supplements and nutraceuticals in the global economy. In: *Dietary Supplements and Nutraceuticals*. Springer Nature Singapore, Singapore, pp. 1–48. DOI: 10.1007/978-981-97-9936-7_29-1.

Choi, S.J. and Choy, J.H. (2014) Biokinetics of zinc oxide nanoparticles: Toxicokinetics, biological fates, and protein interaction. *International Journal of Nanomedicine* 9(sup2), 261–269.

Chong, C.H., Wahi, R., Choo, C.M., Chew, S.J. and Dassanayake, M.K. (2023) *Drying and Valorisation of Food Processing Waste*. CRC Press. DOI: 10.1201/9781003312802.

Chudasama, M. and Goyary, J. (2024) Nanostructured materials in food science: Current progress and future prospects. *Next Materials* 100206. DOI: 10.1016/j.nxmate.2024.100206.

Coupland, J.N. and Hayes, J.E. (2014) Physical approaches to masking bitter taste: Lessons from food and pharmaceuticals. *Pharmaceutical Research* 31(11), 2921–2939. DOI: 10.1007/s11095-014-1480-6.

Cunha, C.N., dos Anjos Pinto, C.B., da Silva Sad, M.E.L., Colturato, V.M.M., Berto, G.L. *et al.* (2024) Influence of micro and nanofibrillated cellulose on sweetened yogurt production and storage. *International Dairy Journal* 155, 105958. DOI: 10.1016/j.idairyj.2024.105958.

Dallaev, R., Papež, N., Allaham, M.M. and Holcman, V. (2025) Biodegradable polymers: Properties, applications, and environmental impact. *Polymers* 17, 14. DOI: 10.3390/polym17141981.

Dam, T.H.T., Grundmann, P., Orozco, R. and Reyhani, N. (2025) Stakeholder-driven circular economy on marginal lands: Governance and sustainability trade-offs in biomass value chains. *Sustainable Production and Consumption*. DOI: 10.1016/j.spc.2025.08.005.

Daramola, O.B., Omole, R.K. and Akinsanola, B.A. (2025) Emerging applications of biorecognition elements-based optical biosensors for food safety monitoring. *Discover Sensors* 1(1), 3.

Das, S., Ghosh, A. and Mukherjee, A. (2021) Nano-encapsulation-based edible coating of essential oils as a novel green strategy against fungal spoilage, mycotoxin contamination, and quality deterioration of stored fruits: An overview. *Frontiers in Microbiology* 12, 768414. DOI: 10.3389/fmicb.2021.768414.

Dasgupta, N., Ranjan, S., Dasgupta, N. and Ranjan, S. (2018) Nanotechnology in food sector. *An Introduction to Food Grade Nano-emulsions* 1–18. DOI: 10.1007/978-981-10-6986-4_1.

Dasgupta, P. and Subramani, D. (2025) Advances in bio-based barrier materials for active food packaging. In: *Green Materials for Active Food Packaging*. Springer Nature, Singapore, pp. 293–333. DOI: 10.1007/978-981-96-0369-5_9.

de Oliveira Mallia, J., Galea, R., Nag, R., Cummins, E., Gatt, R. *et al.* (2022) Nanoparticle food applications and their toxicity: Current trends and needs in risk assessment strategies. *Journal of Food Protection* 85(2), 355–372. DOI: 10.4315/JFP-21-184.

de Sousa, M.S., Schlogl, A.E., Estanislau, F.R., Souza, V.G.L., dos Reis Coimbra, J.S. *et al.* (2023) Nanotechnology in packaging for the food industry: Past, present, and future. *Coatings* 13(8), 1411. DOI: 10.3390/coatings13081411.

del Rosario Herrera-Rivera, M., Torres-Arellanes, S.P., Cortés-Martínez, C.I., Navarro-Ibarra, D.C., Hernández-Sánchez, L. *et al.* (2024) Nanotechnology in food packaging materials: Role and application of nanoparticles. *RSC Advances* 14(30), 21832–21858. DOI: 10.1039/D4RA03711A.

Deshmukh, R.K., Hakim, L., Akhila, K., Ramakanth, D. and Gaikwad, K.K. (2023) Nano clays and its composites for food packaging applications. *International Nano Letters* 13(2), 131–153. DOI: 10.1007/s40089-022-00388-8.

Dhal, S.B. and Kar, D. (2025) Leveraging artificial intelligence and advanced food processing techniques for enhanced food safety, quality, and security: A comprehensive review. *Discover Applied Sciences* 7(1), 1–46. DOI: 10.1007/s42452-025-06472-w.

Dhiman, S., Thakur, B., Kaur, S., Ahuja, M., Gantayat, S. *et al.* (2025) Closing the loop: Technological innovations in food waste valorisation for global sustainability. *Discover Sustainability* 6(1), 1–35. DOI: 10.1007/s43621-025-01073-4.

Diallo, M. and Brinker, C.J. (2011) Nanotechnology for sustainability: Environment, water, food, minerals, and climate. In: *Nanotechnology research directions for societal needs in 2020*. Springer, Dordrecht, pp. 221–259. DOI: 10.1007/978-94-007-1168-6_6

Doa, U.S. (2003) *Nanoscale Science and Engineering for Agriculture and Food Systems: A Report Submitted to Cooperative State Research, Education and Extension Service*. Department of Agriculture, Washington, DC. DOI: 10.1089/ind.2013.1555.

Dobrucka, R. (2016) The future of the active and intelligent packaging industry. *Logforum* 9(2), 103. Available at: https://jbc.bj.uj.edu.pl/Content/936520/NDIGOC102679_9_2013.pdf#page=95

Dong, L., Chen, G., Liu, G., Huang, X., Xu, X. *et al.* (2024) A review on recent advances in the applications of composite Fe3O4 magnetic nanoparticles in the food industry. *Critical Reviews in Food Science and Nutrition* 64(4), 1110–1138. DOI: 10.1080/10408398.2022.2113363.

Dorier, M., Béal, D., Marie-Desvergne, C., Dubosson, M., Barreau, F. *et al.* (2017) Continuous *in vitro* exposure of intestinal epithelial cells to E171 food additive causes oxidative stress, inducing oxidation of DNA bases but no endoplasmic reticulum stress. *Nanotoxicology* 11(6), 751–761. DOI: 10.1080/17435390.2017.1349203.

Duan, K., Pang, G., Duan, Y., Onyeaka, H. and Krebs, J. (2025) Current research development on food contaminants, future risks, regulatory regime and detection technologies: A systematic literature review. *Journal of Environmental Management* 381, 125246.

Dube, E. and Okuthe, G.E. (2025) Silver nanoparticle-based antimicrobial coatings: Sustainable strategies for microbial contamination control. *Microbiology Research* 16(6), 110. DOI: 10.3390/microbiolres16060110.

Durán, N. and Marcato, P.D. (2013) Nanobiotechnology perspectives. Role of nanotechnology in the food industry: A review. *International Journal of Food Science and Technology* 48(6), 1127–1134. DOI: 10.1111/ijfs.12027.

Dusinska, M., Tulinska, J., El Yamani, N., Kuricova, M., Liskova, A. *et al.* (2017) Immunotoxicity, genotoxicity and epigenetic toxicity of nanomaterials: New strategies for toxicity testing. *Food and Chemical Toxicology* 109, 797–811.

Edo, G.I., Mafe, A.N., Ali, A., Akpoghelie, P.O., Yousif, E. *et al.* (2025) Advancing sustainable food packaging: The role of green nanomaterials in enhancing barrier properties. *Food Engineering Reviews* 1–35. DOI: 10.1007/s12393-025-09407-8.

EFSA Scientific Committee, Hardy, A., Benford, D., Halldorsson, T., Jeger, M.J. *et al.* (2018) Guidance on risk assessment of the application of nanoscience and nanotechnologies in the food and feed chain: Part 1, human and animal health. *EFSA Journal* 16(7), e05327. DOI: 10.2903/j.efsa.2018.5327.

Elemike, E.E., Uzoh, I.M., Onwudiwe, D.C. and Babalola, O.O. (2019) The role of nanotechnology in the fortification of plant nutrients and improvement of crop production. *Applied Sciences* 9(3), 499. DOI: 10.3390/app9030499.

Elrashidi, A., Traversa, E. and Elzein, B. (2022) Highly sensitive ultra-thin optical CO2 gas sensors using nanowall honeycomb structure and plasmonic nanoparticles. *Frontiers in Energy Research* 10, 909950.

El-Saadony, M.T., Saad, A.M., Mohammed, D.D.M., Alkafaas, S.S., Ghosh, S. *et al.* (2025) Curcumin, an active component of turmeric: Biological activities, nutritional aspects, immunological, bioavailability, and human health benefits-a comprehensive review. *Frontiers in Immunology* 16, 1603018. DOI: 10.3389/fimmu.2025.1603018.

Elzein, B. (2024) Nano revolution: "Tiny tech, big impact: How nanotechnology is driving SDGs progress". *Heliyon* 10, 10. DOI: 10.1016/j.heliyon.2024.e31393.

Emamifar, A., Kadivar, M., Shahedi, M. and Soleimanian-Zad, S. (2010) Evaluation of nanocomposite packaging containing Ag and ZnO on shelf life of fresh orange juice. *Innovative Food Science & Emerging Technologies* 11(4), 742–748. DOI: 10.1016/j.ifset.2010.06.003.

Emmanuel, O.K., Aria, J., Jose, D. and Diego, C. (2025) Nanotechnology in Food Preservation: Enhancing Shelf Life and Reducing Waste. Available at: https://www.researchgate.net/profile/Emmanuel-Ok-2/publication/389033319

Espitia, P.J.P., Soares, N.D.F.F., Coimbra, J.S.D.R., de Andrade, N.J., Cruz, R.S. *et al.* (2012) Zinc oxide nanoparticles: Synthesis, antimicrobial activity and food packaging applications. *Food and Bioprocess Technology* 5(5), 1447–1464. DOI: 10.1016/B978-0-323-90747-7.00038-7.

Espitia, P.J.P., Otoni, C.G. and Soares, N.F.F. (2025) Zinc oxide nanoparticles for food packaging applications. In: *Antimicrobial Food Packaging*. Academic Press, pp. 603–610. DOI: 10.1016/B978-0-323-90747-7.00038-7.

Ezhilarasi, P.N., Karthik, P., Chhanwal, N. and Anandharamakrishnan, C. (2013) Nano-encapsulation techniques for food bioactive components: A review. *Food and Bioprocess Technology* 6, 628–647. DOI: 10.1007/s11947-012-0944-0.

Fagbohun, T.R., Nji, Q.N., Okechukwu, V.O., Adelusi, O.A., Nyathi, L.A. *et al.* (2025) Aflatoxin exposure in immunocompromised patients: Current state and future perspectives. *Toxins* 17(8), 414.

FAO (2011) *Global Food Losses and Food Waste—Extent, Causes and Prevention*. FAO, Rome.
FAO (2019) Food Wastage Footprint: Impacts on Natural Resources, Summary Report.
FAO (2022) Agricultural production statistics. (2000–2021). FAOSTAT Analytical Brief Series No.60. Rome: FAO. DOI: 10.4060/cc3751en.
Ferdeş, M., Zăbavă, B. Ş., Paraschiv, G., Ionescu, M., Dincă, M.N. *et al.* (2022) Food waste management for biogas production in the context of sustainable development. *Energies* 15(17), 6268.
Ferdous, Z. and Nemmar, A. (2020) Health impact of silver nanoparticles: A review of the biodistribution and toxicity following various routes of exposure. *International Journal of Molecular Sciences* 21(7), 2375. DOI: 10.3390/ijms21072375.
Finglas, P.M., Yada, R.Y. and Toldrá, F. (2014) Nanotechnology in foods: Science behind and future perspectives. *Trends in Food Science & Technology* 40(2), 125–241. DOI: 10.1016/j.tifs.2014.11.001.
Fodor, M., Matkovits, A., Benes, E.L. and Jókai, Z. (2024) The role of near-infrared spectroscopy in food quality assurance: A review of the past two decades. *Foods* 13(21), 3501. DOI: 10.3390/foods13213501.
Food and Agriculture Organization of the United Nations – FAO (2019) Save food: Global initiative on food loss and waste reduction. Available at: http://www.fao.org/save -food/it/ (accessed 16 February 2019).
Fruijtier-Pölloth, C. (2016) The safety of nanostructured synthetic amorphous silica (SAS) as a food additive (E 551). *Archives of Toxicology* 90(12), 2885–2916. DOI: 10.1007/s00204-016-1850-4.
Fuertes, G., Soto, I., Carrasco, R., Vargas, M., Sabattin, J. *et al.* (2016a) Intelligent packaging systems: Sensors and nanosensors to monitor food quality and safety. *Journal of Sensors* 2016(1), 4046061. DOI: 10.1155/2016/4046061.
Fuertes, G., Soto, I., Vargas, M., Valencia, A., Sabattin, J. *et al.* (2016b) Nanosensors for a monitoring system in intelligent and active packaging. *Journal of Sensors* 2016(1), 7980476. DOI: 10.1155/2016/7980476.
Fujihara, J. and Nishimoto, N. (2024) Review of zinc oxide nanoparticles: Toxicokinetics, tissue distribution for various exposure routes, toxicological effects, toxicity mechanism in mammals, and an approach for toxicity reduction. *Biological Trace Element Research* 202(1), 9–23. DOI: 10.1007/s12011-023-03644-w.
Galli, F., Cavicchi, A. and Brunori, G. (2019) Food waste reduction and food poverty alleviation: A system dynamics conceptual model. *Agriculture and Human Values* 36, 289–300. DOI: 10.1007/s10460-019-09919-0.
Gao, X. and Lowry, G.V. (2018) Progress towards standardized and validated characterizations for measuring physicochemical properties of manufactured nanomaterials relevant to nano health and safety risks. *Nano-impact* 9, 14–30. DOI: 10.1016/j.impact.2017.09.002.
Gatoo, M.A., Naseem, S., Arfat, M.Y., Mahmood Dar, A., Qasim, K. *et al.* (2014) Physicochemical properties of nanomaterials: Implication in associated toxic manifestations. *BioMed Research International* 2014(1), 498420.
Gautam, A.K. and Avasthi, S. (2020) Myconanoparticles as potential pest control agents. In: *Nanotechnology for Food, Agriculture, and Environment*. Springer International Publishing, Cham, pp. 189–226. DOI: 10.1007/978-3-030-31938-0_10.
George, J., Aaliya, B., Sunooj, K.V. and Kumar, R. (2022) An overview of higher barrier packaging using nano-additives. *Nanotechnology-Enhanced Food Packaging* 235–264. DOI: 10.1002/9783527827718.ch10.
Ghaani, M., Azimzadeh, M., Buyuktas, D., Carullo, D. and Farris, S. (2024) Electrochemical sensors in the food sector: A review. *Journal of Agricultural and Food Chemistry* 72(44), 24170–24190. DOI: 10.1021/acs.jafc.4c09423.
Ghosh, T., Raj, G.B. and Dash, K.K. (2022) A comprehensive review on nanotechnology based sensors for monitoring quality and shelf life of food products. *Measurement: Food* 7, 100049. DOI: 10.1016/j.meafoo.2022.100049.
Ghosh, A.K., Ghosh, A. and Das, P.K. (2024) Nanotechnology meets stem cell therapy for treating glioblastomas: A review. *ACS Applied Nano Materials* 7(3), 2430–2460. DOI: 10.1021/acsanm.3c04714.
Ghosh, S., Mandal, R.K., Mukherjee, A. and Roy, S. (2025) Nanotechnology in the manufacturing of sustainable food packaging: A review. *Discover Nano* 20(1), 36. DOI: 10.1186/s11671-025-04213-x.

Gibb, M., Pradhan, S.H., Mulenos, M.R., Lujan, H., Liu, J. *et al.* (2021) Characterization of a human *in vitro* intestinal model for the hazard assessment of nanomaterials used in cancer immunotherapy. *Applied Sciences* 11(5), 2113.

Godfray, H.C.J., Poore, J. and Ritchie, H. (2024) Opportunities to produce food from substantially less land. *BMC Biology* 22(1), 138. DOI: 10.1186/s12915-024-01936-8.

González Alejandre, M., Agarwal, V., Martínez Trujillo, M., González Cortes, J.C. and Dasgupta-Schubert, N. (2021) Nanomaterial-aided seed regeneration in the global warming scenario: Multiwalled carbon nanotubes, gold nanoparticles and heat-aged maize seeds. *Applied Nanoscience* 11(5), 1531–1547. DOI: 10.1007/s13204-021-01804-9.

Gu, B., Ju, X., Chang, S.X., Ge, Y. and Chang, J. (2017) Nitrogen use efficiencies in Chinese agricultural systems and implications for food security and environmental protection. *Regional Environmental Change* 17, 1217–1227. DOI: 10.1007/s10113-016-1101-5.

Gul, M., Kashif, M., Muhammad, S., Azizi, S. and Sun, H. (2025) Various methods of synthesis and applications of gold-based nanomaterials: A detailed review. *Crystal Growth & Design* 25(7), 2227–2266.

Guleria, G., Thakur, S., Shandilya, M., Sharma, S., Thakur, S. *et al.* (2023) Nanotechnology for sustainable agro-food systems: The need and role of nanoparticles in protecting plants and improving crop productivity. *Plant Physiology and Biochemistry* 194, 533–549. DOI: 10.1016/j.plaphy.2022.12.004.

Gupta, R.K., Guha, P. and Srivastav, P.P. (2024a) Investigating the toxicological effects of nanomaterials in food packaging associated with human health and the environment. *Journal of Hazardous Materials Letters* 5, 100125. DOI: 10.1016/j.hazl.2024.100125.

Gupta, K., Ashaq, M., Sharma, R., Dutta, A., Panotra, N. *et al.* (2024b) The application of nanotechnology in various food categories and packaging. *Archives of Current Research International* 24(10), 300–307. DOI: 10.9734/acri/2024/v24i10933.

Gupta, P., Mishra, K., Mittal, A.K., Handa, N. and Paul, M.K. (2024c) Current expansion of silver and gold nanomaterials towards cancer theranostics: Development of therapeutics. *Current Nanoscience* 20(3), 356–372. DOI: 10.2174/1573413719666230503144904.

Gupta, R.K., Abd El Gawad, F., Ali, E.A., Karunanithi, S., Yugiani, P. *et al.* (2024d) Nanotechnology: Current applications and future scope in food packaging systems. *Measurement: Food* 13, 100131. DOI: 10.1016/j.meafoo.2023.100131.

Gupta, S., Kumar, D., AE AbdelRahman, M., Mustafa, A.R.A., Scopa, A. *et al.* (2024e) Nano-ecology: Exploring engineered nanoparticles' impact on soil ecosystem health and biodiversity. *Egyptian Journal of Soil Science* 64(4), 1637–1655. DOI: 10.21608/ejss.2024.304704.1814.

Guruprasath, N., Sankarganesh, P., Adeyeye, S.A.O., Babu, A.S. and Parthasarathy, V. (2024) Review on emerging applications of nanobiosensor in food safety. *Journal of Food Science* 89(7), 3950–3972. DOI: 10.1111/1750-3841.17149.

Halonen, N., Pálvölgyi, P.S., Bassani, A., Fiorentini, C., Nair, R. *et al.* (2020) Bio-based smart materials for food packaging and sensors–a review. *Frontiers in Materials* 7, 82. DOI: 10.3389/fmats.2020.00082.

Hanan, E., Dar, A.H., Shams, R. and Goksen, G. (2024) New insights into essential oil nano emulsions loaded natural biopolymers recent development, formulation, characterization and packaging applications: A comprehensive review. *International Journal of Biological Macromolecules* 280, 135751.

Harikrishna, S., Anil, P.P., Shams, R. and Dash, K.K. (2023) Cold plasma as an emerging nonthermal technology for food processing: A comprehensive review. *Journal of Agriculture and Food Research* 14, 100747. DOI: 10.1016/j.jafr.2023.100747.

Hashim, N., Abdullah, S. and Yusoh, K. (2022) Graphene nanomaterials in the food industries: Quality control in promising food safety to consumers. *Graphene and 2D Materials* 7(1), 1–29. DOI: 10.1007/s41127-021-00045-5.

Hassan, N.F., Khattab, T.A., Fouda, M.M., Zaid, A.S.A. and Aboshanab, K.M. (2024) Electrospun cellulose nanofibers immobilized with anthocyanin extract for colorimetric determination of bacteria. *International Journal of Biological Macromolecules* 257, 128817. DOI: 10.1016/j.ijbiomac.2023.128817.

He, X., Deng, H. and Hwang, H.M. (2019) The current application of nanotechnology in food and agriculture. *Journal of Food and Drug Analysis* 27(1), 1–21. DOI: 10.1016/j.jfda.2018.12.002.

Heydari, M. (2024) Cultivating sustainable global food supply chains: A multifaceted approach to mitigating food loss and waste for climate resilience. *Journal of Cleaner Production* 442, 141037. DOI: 10.1016/j.jclepro.2024.141037.

Hiywotu, A.M. (2025) Advancing sustainable agriculture for goal 2: Zero hunger-a comprehensive overview of practices, policies, and technologies. *Agroecology and Sustainable Food Systems* 1–29. DOI: 10.1080/21683565.2025.2451344.

Hofer, S., Hofstätter, N., Punz, B., Hasenkopf, I., Johnson, L. *et al.* (2022) Immunotoxicity of nanomaterials in health and disease: Current challenges and emerging approaches for identifying immune modifiers in susceptible populations. *Wiley Interdisciplinary Reviews: Nanomedicine and Nanobiotechnology* 14(6), e1804. DOI: 10.1002/wnan.1804.

Hossain, A., Skalicky, M., Brestic, M., Mahari, S., Kerry, R.G. *et al.* (2021) Application of nanomaterials to ensure quality and nutritional safety of food. *Journal of Nanomaterials* 2021(1), 9336082. DOI: 10.1155/2021/9336082.

Hu, Y., Zhao, J. and Li, Q. (2024) Internet of things-based food safety monitoring: Integration of smart sensors and AI analytics. *Food Control* 158, 109546. DOI: 10.1016/j.foodcont.2023.109546.

Huang, W. and Wang, X. (2024) The impact of technological innovations on agricultural productivity and environmental sustainability in China. *Sustainability* 16(19), 8480. DOI: 10.3390/su16198480.

Hussain, S., Akhter, R. and Maktedar, S.S. (2024) Advancements in sustainable food packaging: From eco-friendly materials to innovative technologies. *Sustainable Food Technology* 2(5), 1297–1364. DOI: 10.1039/D4FB00084F.

Inês, A. and Cosme, F. (2025) Biosensors for detecting food contaminants—An overview. *Processes* 13(2). DOI: 10.3390/pr13020380.

Jafarizadeh-Malmiri, H., Sayyar, Z., Anarjan, N., Berenjian, A., Jafarizadeh-Malmiri, H. *et al.* (2019) Nano-sensors in food nanobiotechnology. *Nanobiotechnology in Food: Concepts, Applications and Perspectives* 81–94. DOI: 10.1007/978-3-030-05846-3_6.

Jahangiri, F., Mohanty, A.K. and Misra, M. (2024) Sustainable biodegradable coatings for food packaging: Challenges and opportunities. *Green Chemistry* 26(9), 4934–4974. DOI: 10.1039/D3GC02647G.

Janik-Karpinska, E., Ceremuga, M., Niemcewicz, M., Podogrocki, M., Stela, M. *et al.* (2022) Immunosensors—the future of pathogen real-time detection. *Sensors* 22(24), 9757. DOI: 10.3390/s22249757.

Jayan, H., Zhou, R., Sun, C., Wang, C., Yin, L. *et al.* (2025) Intelligent gas sensors for food safety and quality monitoring: Advances, applications, and future directions. *Foods* 14(15), 2706. DOI: 10.3390/foods14152706.

Jia, X., Wen, Q., Sun, Y., Chen, Y., Gao, D. *et al.* (2024) Preparation and performance of PBAT/PLA/CaCO3 composites via solid-state shear milling technology. *Polymers* 16(20), 2942.

Jin, M., Li, N., Sheng, W., Ji, X., Liang, X. *et al.* (2021) Toxicity of different zinc oxide nanomaterials and dose-dependent onset and development of Parkinson's disease-like symptoms induced by zinc oxide nanorods. *Environment International* 146, 106179. DOI: 10.1016/j.envint.2020.106179.

Kakimova, Z., Orynbekov, D., Zharykbasova, K., Kakimov, A., Zharykbasov, Y. *et al.* (2023) Advancements in nano bio sensors for food quality and safety assurance-a review. *Slovak Journal of Food Sciences/Potravinarstvo* 17(1). DOI: 10.5219/1903.

Kang, S., Hao, X., Du, T., Tong, L., Su, X. *et al.* (2017) Improving agricultural water productivity to ensure food security in China under changing environment: From research to practice. *Agricultural Water Management* 179, 5–17. DOI: 10.1016/j.agwat.2016.05.007.

Karimi, M., Sahandi Zangabad, P., Ghasemi, A., Amiri, M., Bahrami, M. *et al.* (2016) Temperature-responsive smart nanocarriers for delivery of therapeutic agents: Applications and recent advances. *ACS Applied Materials & Interfaces* 8(33), 21107–21133. DOI: 10.1021/acsami.6b00371.

Karnwal, A., Kumar, G., Singh, R., Selvaraj, M., Malik, T. *et al.* (2025a) Natural biopolymers in edible coatings: Applications in food preservation. *Food Chemistry* X, 102171. DOI: 10.1016/j.fochx.2025.102171.

Karnwal, A., Rauf, A., Jassim, A.Y., Selvaraj, M., Al-Tawaha, A.R.M.S. *et al.* (2025b) Advanced starch-based films for food packaging: Innovations in sustainability and functional properties. *Food Chemistry* X, 102662. DOI: 10.1016/j.fochx.2025.102662.

Khan, Y., Sadia, H., Ali Shah, S.Z., Khan, M.N., Shah, A.A. *et al.* (2022) Classification, synthetic, and characterization approaches to nanoparticles, and their applications in various fields of nanotechnology: A review. *Catalysts* 12(11), 1386.

Khan, W.S., Asmatulu, E. and Asmatulu, R. (2025) Nanotechnology emerging trends, markets and concerns. In: *Nanotechnology Safet.* Elsevier, pp. 1–21. DOI: 10.1016/B978-0-443-15904-6.00017-4.

Khorshidi, S., Mehdizadeh, T. and Ghorbani, M. (2021) The effect of chitosan coatings enriched with the extracts and essential oils of *Elettaria cardamomum* on the shelf-life of chicken drumsticks vacuum-packaged at 4 C. *Journal of Food Science and Technology* 58(8), 2924–2935. DOI: 10.1007/s13197-020-04794-8.

Kilemile, W., Vulla, K.E., Mihafu, F. and Chandrasekaran, V. (2025) Transforming food systems: A review of sustainable approaches to minimize food loss and waste. *Food Science & Nutrition* 13(11), e71167. DOI: 10.1002/fsn3.71167.

Kim, T., Lee, M., Ko, H. and Cho, D.I. (2024) Piezoresistance of silicon nanowires for sensing applications: Optimizing nanowire parameters from electrical and mechanical perspectives. *IEEE Access*. DOI: 10.1109/ACCESS.2024.3427823.

Kingra, P.K. and Misra, A.K. (2022) Agricultural input use efficiency and climate change: Ways to improve the environment and food security. In: *Input Use Efficiency for Food and Environmental Security*. Springer Nature Singapore, Singapore, pp. 33–67. DOI: 10.1007/978-981-16-5199-1_2.

Konfo, T.R.C., Tchekessi, C.K.C. and Baba-Moussa, F.A.K. (2024) Status report on innovations and applications of smart bio-systems for real-time monitoring of food quality. *Applied Food Research* 100546. DOI: 10.1016/j.afres.2024.100546.

Kuang, L., Burgess, B., Cuite, C.L., Tepper, B.J. and Hallman, W.K. (2020) Sensory acceptability and willingness to buy foods presented as having benefits achieved through the use of nanotechnology. *Food Quality and Preference* 83, 103922. Available at: https://www.sciencedirect.com/science/article/pii/S0950329319306081

Kulkarni, M.B., Ayachit, N.H. and Aminabhavi, T.M. (2022) Recent advancements in nanobiosensors: Current trends, challenges, applications, and future scope. *Biosensors* 12(10), 892. DOI: 10.3390/bios12100892.

Kumar, S. (2024) Prospects and challenges of nanomaterials in sustainable food preservation and packaging: A review. *Discover Nano* 19(1), 1–18. DOI: 10.1186/s11671-024-04142-1.

Kumar, D., Farrukh, M. and Faisal, N. (2021a) Nanocomposites in the food packaging industry: Recent trends and applications. *Research Anthology on Food Waste Reduction and Alternative Diets for Food and Nutrition Security* 122–146. DOI: 10.4018/978-1-7998-5354-1.ch006.

Kumar, P., Pathak, D. and Singh, P. (2021b) Nanostructured materials for food packaging and sensor applications: A review. *Critical Reviews in Food Science and Nutrition* 61(4), 561–583. DOI: 10.1039/D5FB00379B.

Kumar, C.V., Karthick, V., Kumar, V.G., Inbakandan, D., Rene, E.R. *et al.* (2022) The impact of engineered nanomaterials on the environment: Release mechanism, toxicity, transformation, and remediation. *Environmental Research* 212, 113202. DOI: 10.1016/j.envres.2022.113202.

Kumar, S., Sidhu, H.K., Paul, A.K., Bhardwaj, N., Thakur, N.S. *et al.* (2023) Bioengineered multi-walled carbon nanotube (MWCNT) based biosensors and applications thereof. *Sensors & Diagnostics* 2(6), 1390–1413. DOI: 10.1039/D3SD00176H.

Kumar, D., Kumari, A., Sati, P., Kumar, S. and Verma, A.S. (2024) Role of nanotechnology in food processing. *Nonthermal Food Processing, Safety, and Preservation* 153–169.

Kumar, S., Wang, S. and Keerio, H.A. (2025) Application of Fe-NPs for enhancing methane production in an anaerobic biogas plant. *ACS Omega* 10(39), 44783–44797.

Kumari, A. and Chauhan, A.K. (2022) Iron nanoparticles as a promising compound for food fortification in iron deficiency anemia: A review. *Journal of Food Science and Technology* 59(9), 3319–3335. DOI: 10.1007/s13197-021-05184-4.

Kumari, A., Vyas, V. and Kumar, S. (2022) Synthesis, characterization, and applications of gold nanoparticles in development of plasmonic optical fiber-based sensors. *Nanotechnology* 34(4), 042001. DOI: 10.1088/1361-6528/ac9982.

Kuppusamy, S., Meivelu, M., Praburaman, L., Alam, M.M. and Al-Sehemi, A.G. (2024) Integrating AI in food contaminant analysis: Enhancing quality and environmental protection. *Journal of Hazardous Materials Advances* 16, 100509. DOI: 10.1016/j.hazadv.2024.100509.

Kusuma, S.A.F., Harmonis, J.A., Pratiwi, R. and Hasanah, A.N. (2023) Gold nanoparticle-based colorimetric sensors: Properties and application in detection of heavy metals and biological molecules. *Sensors* 23(19), 8172. DOI: 10.3390/s23198172.

Kwon, S. and Ko, S. (2022) Colorimetric freshness indicator based on cellulose nanocrystal–silver nanoparticle composite for intelligent food packaging. *Polymers* 14(17), 3695. DOI: 10.3390/polym14173695.

Lei, Q., Zou, Y., Yan, J., Cheng, J., Guo, J. *et al.* (2025) Advances and challenges in the application of artificial intelligence to medical biosensing technology: A review. *TrAC Trends in Analytical Chemistry* 118592.

Leslie, J.F., Ezekiel, C.N., Wagner, M., Elliott, C., McNerney, O. *et al.* (2025) A FoodSafeR perspective on emerging food safety hazards and associated risks. *Frontiers in Sustainable Food Systems* 9, 1646792. DOI: 10.3389/fsufs.2025.1646792.

Li, M., Zhang, Y., Feng, S., Zhang, X., Xi, Y. *et al.* (2022) Bioaccumulation and biomagnification effects of nano-TiO2 in the aquatic food chain. *Ecotoxicology* 31(6), 1023–1034. DOI: 10.1007/s10646-022-02572-0.

Li, Y., Feng, X., Huai, Y., Hassan, M.U., Cui, Z. *et al.* (2024) Enhancing crop productivity and resilience by promoting soil organic carbon and moisture in wheat and maize rotation. *Agriculture, Ecosystems & Environment* 368, 109021. DOI: 10.1016/j.agee.2024.109021.

Liao, W., Badri, W., Dumas, E., Ghnimi, S., Elaissari, A. *et al.* (2021) Nano-encapsulation of essential oils as natural food antimicrobial agents: An overview. *Applied Sciences* 11(13), 5778. DOI: 10.3390/app11135778.

Lisboa, H.M., Pasquali, M.B., dos Anjos, A.I., Sarinho, A.M., de Melo, E.D. *et al.* (2024) Innovative and sustainable food preservation techniques: Enhancing food quality, safety, and environmental sustainability. *Sustainability* 16(18), 8223. DOI: 10.3390/su16188223.

Liu, X., Liu, W., Ren, Z., Ma, Y., Dong, B. *et al.* (2021) Progress of optomechanical micro/nano sensors: A review. *International Journal of Optomechatronics* 15(1), 120–159. DOI: 10.1080/15599612.2021.1986612.

Lopez, R.M., White, J.R., Truong, L. and Tanguay, R.L. (2022) Size-and oxidation-dependent toxicity of graphene oxide nanomaterials in embryonic zebrafish. *Nanomaterials* 12(7), 1050. DOI: 10.3390/nano12071050.

Lu, H., Zhang, S., Wang, J. and Chen, Q. (2021) A review on polymer and lipid-based nanocarriers and its application to nano-pharmaceutical and food-based systems. *Frontiers in Nutrition* 8, 783831. DOI: 10.3389/fnut.2021.783831.

Ma, M., Yang, X., Ying, X., Shi, C., Jia, Z. *et al.* (2023) Applications of gas sensing in food quality detection: A review. *Foods* 12(21), 3966. DOI: 10.3390/foods12213966.

Magnuson, B.A., Jonaitis, T.S. and Card, J.W. (2011) A brief review of the occurrence, use, and safety of food-related nanomaterials. *Journal of Food Science* 76(6), R126–R133. DOI: 10.1111/j.1750-3841.2011.02170.x.

Manzoor, S., Fayaz, U., Dar, A.H., Dash, K.K., Shams, R. *et al.* (2024) Sustainable development goals through reducing food loss and food waste: A comprehensive review. *Future Foods* 9, 100362.

Marta, H., Wijaya, C., Sukri, N., Cahyana, Y. and Mohammad, M. (2022) A comprehensive study on starch nanoparticle potential as a reinforcing material in bioplastic. *Polymers* 14(22), 4875.

Mathur, S., Pareek, S. and Shrivastava, D. (2022) Nanofertilizers for development of sustainable agriculture. *Communications in Soil Science and Plant Analysis* 53(16), 1999–2016. DOI: 10.1080/00103624.2022.2070191.

Maurya, A., Singh, V.K., Das, S., Prasad, J., Kedia, A. *et al.* (2021) Essential oil nano-emulsion as eco-friendly and safe preservative: Bioefficacy against microbial food deterioration and toxin secretion, mode of action, and future opportunities. *Frontiers in Microbiology* 12, 751062.

Maurya, V.K., Shakya, A., McClements, D.J., Srinivasan, R., Bashir, K. *et al.* (2023) Vitamin C fortification: Need and recent trends in encapsulation technologies. *Frontiers in Nutrition* 10, 1229243. DOI: 10.3389/fnut.2023.1229243.

Medina-Ramirez, I.E., Macías-Díaz, J.E., Masuoka-Ito, D. and Zapien, J.A. (2024) Holotomography and atomic force microscopy: A powerful combination to enhance cancer, microbiology and nanotoxicology research. *Discover Nano* 19(1), 64. DOI: 10.1186/s11671-024-04003-x.

Miah, M.S., Chy, M.W.R., Ahmed, T., Suchi, M., Muhtady, M.A. *et al.* (2025) Emerging trends in nanotechnologies for vitamin delivery: Innovation and future prospects. *Nano Trends* 100126. DOI: 10.1016/j.nwnano.2025.100126.

Mirmehdi, S., Hein, P.R.G., de Luca Sarantópoulos, C.I.G., Dias, M.V. and Tonoli, G.H.D. (2018) Cellulose nanofibrils/nanoclay hybrid composite as a paper coating: Effects of spray time, nanoclay content and corona discharge on barrier and mechanical properties of the coated papers. *Food Packaging and Shelf Life* 15, 87–94.

Mishra, R.K., Vinu, R. and Ghosh, S. (2023) Recent developments in nano-biosensors for food safety assessment. *Biosensors and Bioelectronics* 218, 114784. DOI: 10.1016/j.bios.2022.114784.

Mittal, D., Kaur, G., Singh, P., Yadav, K. and Ali, S.A. (2020) Nanoparticle-based sustainable agriculture and food science: Recent advances and future outlook. *Frontiers in Nanotechnology* 2, 579954.

Momirović, N., Nikolić, D., Jakanovski, M.V., Drami, A., Mosić, M.D. *et al.* (2023) The sugars content of parental and new perspective descendant strawberry genotypes potential approach for the future selection process. In: *XXII Congress EuroFoodChem*, Beograd: Srpsko hemijsko društvo, Belgrade, Serbia, June 14–16, 2023, pp. 175–175. Available at: https://cherry.chem.bg.ac.rs/handle/123456789/5974

Monavari, S.M., Marsusi, F., Memarian, N. and Qasemnazhand, M. (2023) Carbon nanotubes and nanobelts as potential materials for biosensor. *Scientific Reports* 13(1), 3118.

Moradi, M., Molaei, R., Kousheh, S.A., T. Guimarães, J. and McClements, D.J. (2023) Carbon dots synthesized from microorganisms and food by-products: Active and smart food packaging applications. *Critical Reviews in Food Science and Nutrition* 63(14), 1943–1959. DOI: 10.1080/10408398.2021.2015283.

Muñoz-Tebar, N., Pérez-Álvarez, J.A., Fernández-López, J. and Viuda-Martos, M. (2023) Chitosan edible films and coatings with added bioactive compounds: Antibacterial and antioxidant properties and their application to food products: A review. *Polymers* 15(2), 396. DOI: 10.3390/polym15020396.

Musa, A.H., Musa, S.S., Alhassan, M.Y., Okesanya, O.J., Ishak, A.S., Auwal, A.R., Alaka, H.O., Othman, Z.K., Ibrahim, A.M., Sy, F.A. and Ahmed, M.M. (2025) Role of nanotechnology in climate change mitigation: Opportunities, challenges, and future directions. *Narra X* 3(3), 212–212. DOI: 10.52225/narrax.v3i3.212.

Mustapha, L.S., Obayomi, O.V., Lau, S.Y. and Obayomi, K.S. (2025) The role of nanotechnology in agricultural systems with emphasis on water management. *Inorganic Chemistry Communications* 115507. DOI: 10.1016/j.inoche.2025.115507.

Muthu, A., Nguyen, D.H., Neji, C., Törős, G., Ferroudj, A. *et al.* (2025) Nanomaterials for smart and sustainable food packaging: Nano-sensing mechanisms, and regulatory perspectives. *Foods* 14(15), 2657. DOI: 10.3390/foods14152657.

Nakagawa, K. (2014) Nano-and microencapsulation of flavor in food systems. *Nano-and Microencapsulation for Foods* 249–271. DOI: 10.1002/9781118292327.ch10.

Nath, R. (2024) Applications of smart sensors in food quality and safety monitoring. *Critical Reviews in Food Science and Nutrition* 64(5), 765–778. DOI: 10.1080/10408398.2022.2098393.

Nath, D., Santhosh, R., Pal, K. and Sarkar, P. (2022) Nanoclay-based active food packaging systems: A review. *Food Packaging and Shelf Life* 31, 100803.

Neethirajan, S., Ragavan, V., Weng, X. and Chand, R. (2018) Biosensors for sustainable food engineering: Challenges and perspectives. *Biosensors* 8(1), 23. DOI: 10.3390/bios8010023.

Negi, R. and Yadav, S. (2025) A comprehensive review on edible packaging and its formation methods with recent eco-friendly advancements. *Biopolymers* 116(6), e70053. DOI: 10.1002/bip.70053.

Nile, S.H., Baskar, V., Selvaraj, D., Nile, A., Xiao, J. *et al.* (2020) Nanotechnologies in food science: Applications, recent trends, and future perspectives. *Nano-Micro Letters* 12(1), 45. DOI: 10.1007/s40820-020-0383-9.

Ninago, M.D., Giaroli, M.C., Passaretti, M.G., Villar, M.A. and López, O.V. (2024) Polymer–clay nanocomposites for food packaging. In: *Nanostructured Materials for Food Packaging Applications*. Elsevier, pp. 189–213.

Ong, K.J., Shatkin, J.A., Nelson, K., Ede, J.D. and Retsina, T. (2017) Establishing the safety of novel bio-based cellulose nanomaterials for commercialization. *Nano-impact* 6, 19–29. DOI: 10.1016/j.impact.2017.03.002.

Onyeaka, H., Passaretti, P., Miri, T. and Al-Sharify, Z.T. (2022) The safety of nanomaterials in food production and packaging. *Current Research in Food Science* 5, 763–774. DOI: 10.1016/j.crfs.2022.04.005.

Oomen, A.G., Steinhäuser, K.G., Bleeker, E.A., van Broekhuizen, F., Sips, A. *et al.* (2018) Risk assessment frameworks for nanomaterials: Scope, link to regulations, applicability, and outline for future directions in view of needed increase in efficiency. *Nano-impact* 9, 1–13. DOI: 10.1016/j.impact.2017.09.001.

Operato, L., Panzeri, A., Masoero, G., Gallo, A., Gomes, L. and Hamd, W. (2025) Food packaging use and post-consumer plastic waste management: A comprehensive review. *Frontiers in Food Science and Technology* 5, 1520532. DOI: 10.3389/frfst.2025.1520532.

Osman, A.I., Zhang, Y., Farghali, M., Rashwan, A.K., Eltaweil, A.S. *et al.* (2024) Synthesis of green nanoparticles for energy, biomedical, environmental, agricultural, and food applications: A review. *Environmental Chemistry Letters* 22(2), 841–887. DOI: 10.1007/s10311-023-01682-3.

Ottaway, P.B. (2009) Nanotechnology in supplements and foods–EU concerns. *Nutraceuticals International* 1.

Ozturk, B., Argin, S., Ozilgen, M. and McClements, D.J. (2015) Formation and stabilization of nano-emulsion-based vitamin E delivery systems using natural biopolymers: Whey protein isolate and gum arabic. *Food Chemistry* 188, 256–263. DOI: 10.1016/j.foodchem.2015.05.005.

Paliwal, T., Kesharwani, P., Yadav, R., Verma, K., Dwivedi, J. *et al.* (2025) Emerging nanomaterial-based sensors for foodborne pathogen and toxin detection. *Food Research International* 117083. DOI: 10.1016/j.foodres.2025.117083.

Panda, H.S., Rout, H.S. and Jakovljevic, M., 2024. Catastrophic health expenditure of inpatients in emerging economies: Evidence from the Indian subcontinent. *Health Research Policy and Systems* 22(1), 104. DOI: 10.1186/s12961-024-01202-x.

Pandey, A., Kamran, F., Choudhury, M., Li, L., Rahman, M.S. *et al.* (2025a) Omega-3 fatty acid fortification of plant-based beverages to enhance their nutritional profile. *Foods* 14(9), 1602. DOI: 10.3390/foods14091602.

Pandey, V.K., Singh, G., Choudhary, P., Pathak, A. and Saini, A. (2025b) Nanosensors in food shelf-life extension and quality monitoring. In: *Advancements in Nanotechnology for Food and Packaging*. Elsevier, pp. 79–102. DOI: 10.1016/B978-0-443-21428-8.00004-7.

Pandhi, S., Mahato, D.K. and Kumar, A. (2023) Overview of green nanofabrication technologies for food quality and safety applications. *Food Reviews International* 39(1), 240–260. DOI: 10.1080/87559129.2021.1904254.

Papadiamantis, A.G., Mavrogiorgis, A., Papatzelos, S., Mintis, D., Melagraki, G. *et al.* (2024) A systematic review on the state-of-the-art and research gaps regarding inorganic and carbon-based multicomponent and high-aspect ratio nanomaterials. *Computational and Structural Biotechnology Journal* 25, 211–229. DOI: 10.1016/j.csbj.2024.10.020.

Passeri, D., Atanasio, P., Proietti, A., Mancini, C., Cognigni, F. *et al.* (2024) Analytical methods for nanomaterials investigation in food and food-related materials. *Applied Food Research* 100675.

Patial, B., Chelak, P., Bansal, A., Gupta, R. and Mittal, S.K. (2025) Challenges and limitations of implementing the nanobiosensors in food sector. In: *Nanobiosensors for the Food Industry*. Springer Nature Singapore, Singapore, pp. 317–330.

Pattnaik, R., Panda, S.K., Biswas, S., De, S., Satahrada, S. *et al.* (2024) Prospects and challenges of nanomaterials in sustainable food preservation and packaging: A review. *Discover Nano* 19(1), 178. DOI: 10.1186/s11671-024-04142-1.

Peng, B., Qi, X., Qiao, L., Lu, J., Qian, Z. *et al.* (2025) Nanocomposite-enabled next-generation food packaging: A comprehensive review on advanced preparation methods, functional properties, preservation applications, and safety considerations. *Functional Properties, Preservation Applications, and Safety Considerations. Foods* 14(21), 3688.

Pengsomjit, U., Alabdo, F., Karuwan, C., Kraiya, C., Alahmad, W. *et al.* (2024) Innovative graphene-based nanocomposites for improvement of electrochemical sensors: Synthesis, characterization, and applications. *Critical Reviews in Analytical Chemistry* 1–19. DOI: 10.1080/10408347.2024.2343854.

Perera, K.Y., Jaiswal, A.K. and Jaiswal, S. (2023) Biopolymer-based sustainable food packaging materials: Challenges, solutions, and applications. *Foods* 12(12), 2422.

Peters, R., Kramer, E., Oomen, A.G., Herrera Rivera, Z.E., Oegema, G. *et al.* (2012) Presence of nano-sized silica during *in vitro* digestion of foods containing silica as a food additive. *ACS Nano* 6(3), 2441–2451. DOI: 10.1021/nn204728k.

Pirozzi, A., Ferrari, G. and Donsì, F. (2021) The use of nanocellulose in edible coatings for the preservation of perishable fruits and vegetables. *Coatings* 11(8), 990. DOI: 10.3390/coatings11080990.

Piryaei, M. and Azimi, S. (2024) Preparation and evaluation of smart food packaging films with anthocyanin Sardasht black grape based on *Astragalus gummifer* and chitosan nanoparticles. *International Journal of Biological Macromolecules* 254, 127974. DOI: 10.1016/j.ijbiomac.2023.127974.

Pradhan, N., Singh, S., Ojha, N., Shrivastava, A., Barla, A. *et al.* (2015) Facets of nanotechnology are seen in the food processing, packaging, and preservation industry. *BioMed Research International* 2015(1), 365672. DOI: 10.1155/2015/365672.

Prakash, J., Cho, J. and Mishra, Y.K. (2022) Photocatalytic TiO2 nanomaterials as potential antimicrobial and antiviral agents: Scope against blocking the SARS-COV-2 spread. *Micro and Nano Engineering* 14, 100100. DOI: 10.1016/j.mne.2021.100100.

Prasad, R.D., Sahoo, A.K., Shrivastav, O.P., Charmode, N., Prasad, S.R. *et al.* (2022) A review on aspects of nanotechnology in food science and animal nutrition. *ES Food & Agroforestry* 8, 12–46. DOI: 10.30919/esfaf704.

Pugazhendhi, A., Alshehri, M.A., Kandasamy, S., Sarangi, P.K. and Sharma, A. (2025) Deciphering the importance of nano-encapsulation to improve the availability of bioactive molecules in food sources to the human body. *Food Chemistry* 464, 141762. DOI: 10.1016/j.foodchem.2024.141762.

Pulikkalparambil, H., Phothisarattana, D., Promhuad, K. and Harnkarnsujarit, N. (2023) Effect of silicon dioxide nanoparticle on microstructure, mechanical and barrier properties of biodegradable PBAT/PBS food packaging. *Food Bioscience* 55, 103023. DOI: 10.1016/j.fbio.2023.103023.

Qadir, S.A. (2025) Emerging trends in nano-sensors: A new frontier in food safety and quality assurance. *Heliyon* 11(2025), e41181. DOI: 10.1016/j.heliyon.2024.e41181.

Qu, L., Zhang, X., Chu, Y., Zhang, Y., Lin, Z. *et al.* (2025) Research progress on nanotechnology-driven enzyme biosensors for electrochemical detection of biological pollution and food contaminants. *Foods* 14(7), 1254. DOI: 10.3390/foods14071254.

Rabiee, N., Sharma, R., Foorginezhad, S., Jouyandeh, M., Asadnia, M. *et al.* (2023) Green and sustainable membranes: A review. *Environmental Research* 231, 116133.

Rahman, B.A., Viphavakit, C., Chitaree, R., Ghosh, S., Pathak, A.K. *et al.* (2022) Optical fiber, nanomaterial, and thz-metasurface-mediated nano-biosensors: A review. *Biosensors* 12(1), 42. DOI: 10.3390/bios12010042.

Rajendran, A.T. and Vadakkepushpakath, A.N. (2024) Natural food components as biocompatible carriers: A novel approach to glioblastoma drug delivery. *Foods* 13(17), 2812. DOI: 10.3390/foods13172812.

Ramesh, M., Janani, R., Deepa, C. and Rajeshkumar, L. (2022) Nanotechnology-enabled biosensors: A review of fundamentals, design principles, materials, and applications. *Biosensors* 13(1), 40. DOI: 10.3390/bios13010040.

Ramezani, G., Assadpour, E., Zhang, W. and Jafari, S.M. (2025) Carbon nanomaterial-based sensors for smart packaging of food products. *Industrial Crops and Products* 232, 121315. DOI: 10.1016/j.indcrop.2025.121315.

Ramezanzadeh, B., Haeri, Z. and Ramezanzadeh, M. (2016) A facile route of making silica nanoparticles-covered graphene oxide nanohybrids (SiO2-GO); fabrication of SiO2-GO/epoxy composite coating with superior barrier and corrosion protection performance. *Chemical Engineering Journal* 303, 511–528. DOI: 10.1016/j.cej.2016.06.028.

Rana, L., Kumar, M., Rajput, J., Kumar, N., Sow, S. *et al.* (2024) Nexus between nanotechnology and agricultural production systems: Challenges and future prospects. *Discover Applied Sciences* 6(11), 555. DOI: 10.1007/s42452-024-06265-7.

Rana, M.S., Khatun, M.A., Anis-Ul-Haque, K.M. and Rahman, M.W. (2025) Nanopaints in food packaging applications. In: *Nanopaints*. Elsevier, pp. 359–387. DOI: 10.1016/B978-0-443-26512-9.00014-8.

Rani, H., Kakkar, P. and Singh, D.P. (2025a) Advancements in precision agriculture for maximizing crop yield and minimizing waste via innovative technological solutions. *International Journal of Research and Review in Applied Science, Humanities, and Technology* 1–8.

Rani, R., Ananna., Sand, A. and Tuteja, J. (2025b) Nanosensors for environmental pollutants. In: *Advances in Nanomaterials for Detection, Control, and Removal of Environmental Pollutants*. Springer Nature Switzerland, Cham, pp. 35–63. DOI: 10.1007/978-3-031-87409-3_2.

Rezagholizade-Shirvan, A., Soltani, M., Shokri, S., Radfar, R., Arab, M. *et al.* (2024) Bioactive compound encapsulation: Characteristics, applications in food systems, and implications for human health. *Food Chemistry: X* 24, 101953. DOI: 10.1016/j.fochx.2024.101953.

Rizvi, S.S., Moraru, C.I., Bouwmeester, H., Kampers, F.W. and Cheng, Y. (2022) Nanotechnology and food safety. In: *Ensuring Global Food Safety*. Academic Press, pp. 325–340. DOI: 10.1016/B978-0-12-816011-4.00016-1.

Rojas, K., Verdugo-Molinares, M.G. and Vallejo-Cardona, A.A. (2024) Use of encapsulating polymers of active compounds in the pharmaceutical and food industry. *Food Chemistry Advances* 4, 100619. DOI: 10.1016/j.focha.2024.100619.

Rouf, A., Mohammed, T., Okunola, A. and Han, E. (2025) Scalable Fabrication Techniques for Quantum Dot-Polymer Sensors.

Roychoudhury, A. (2020) Yeast-mediated green synthesis of nanoparticles for biological applications. *Indian Journal of Pharmaceutical and Biological Research (IJPBR)* (3), 26–31. DOI: 10.30750/ijpbr.8.3.4.

Ryabtseva, M.S., Umanskaya, S.F., Shevchenko, M.A., Krivobok, V.S., Kolobov, A.V. *et al.* (2023) Transformation of nano-size titanium dioxide particles in the gastrointestinal tract and its role in the transfer of nanoparticles through the intestinal barrier. *International Journal of Molecular Sciences* 24(19), 14911.

Sabliov, C., Chen, H. and Yada, R. (eds) (2015) *Nanotechnology and Fnctional Foods: Effective Delivery of Bioactive Ingredients*. John Wiley & Sons.

Sahani, S. and Sharma, Y.C. (2021) Advancements in applications of nanotechnology in global food industry. *Food Chemistry* 342, 128318. DOI: 10.1016/j.foodchem.2020.128318.

Sahoo, M., Vishwakarma, S., Panigrahi, C. and Kumar, J. (2021) Nanotechnology: Current applications and future scope in food. *Food Frontiers* 2(1), 3–22. DOI: 10.1002/fft2.58.

Sapna, 0., Sharma, C., Pathak, P. and Gautam, S. (2025) Chitosan edible coatings loaded with bioactive components for fruits and vegetables: A step toward sustainable development goals. *Food and Bioprocess Technology* 1–35. DOI: 10.1007/s11947-025-03770-4.

Sarabandi, M., Zargar, M., Ghorbani, A. and Chen, M. (2025) Smart and sustainable nano-biosensing technologies for advancing stress detection and management in agriculture and beyond. *Industrial Crops and Products* 226, 120713.

Sarfraz, J., Gulin-Sarfraz, T., Nilsen-Nygaard, J. and Pettersen, M.K. (2020) Nanocomposites for food packaging applications: An overview. *Nanomaterials* 11(1), 10.

Sarode, R.J. and Mahajan, H.S. (2024) Dendrimers for drug delivery: An overview of its classes, synthesis, and applications. *Journal of Drug Delivery Science and Technology* 98, 105896. DOI: 10.1016/j.jddst.2024.105896.

Sathiensathaporn, S., Solé-Porta, A., Baowan, D., Pissuwan, D., Wongtrakoongate, P. *et al.* (2025) Nano-encapsulation of vitamin B2 using chitosan-modified poly (lactic-co-glycolic acid) nanoparticles: Synthesis, characterization, and *in vitro* studies on simulated gastrointestinal stability and delivery. *Journal of Food Science* 90(1), e17631. DOI: 10.1111/1750-3841.17631.

Sati, A., Mali, S.N., Samdani, N., Annadurai, S., Dongre, R. *et al.* (2025) From past to present: Gold nanoparticles (aunps) in daily life— synthesis mechanisms, influencing factors. *Characterization, Toxicity, and Emerging Applications in Biomedicine, Nano-electronics, and Materials Science. ACS Omega* 10(31), 33999–34087.

Scott, N.R., Chen, H. and Cui, H. (2018) Nanotechnology applications and implications of agrochemicals toward sustainable agriculture and food systems. *Journal of Agricultural and Food Chemistry* 66(26), 6451–6456. DOI: 10.1021/acs.jafc.8b00964.

Shaaban, H.A. and Farouk, A. (2022) Preservation and packaging food recent methods and techniques: A review. *Biomedical Journal of Scientific & Technical Research* 46, 37680–37695. DOI: 10.26717/BJSTR.2022.46.007384.

Shafiq, M., Anjum, S., Hano, C., Anjum, I. and Abbasi, B.H. (2020) An overview of the applications of nanomaterials and nanodevices in the food industry. *Foods* 9(2), 148. DOI: 10.3390/foods9020148.

Shah, D.P. and Krishnakumar, A. (2025) Green nanotechnology in food packaging and preservation. In: *Green Nanotechnology Applications for Ecosystem Sustainability*. Springer Nature Switzerland, Cham, pp. 249–305. DOI: 10.1007/978-3-031-90766-1_9.

Shah, M.A., Shahnaz, T., Masoodi, J.H., Nazir, S., Qurashi, A. *et al.* (2024) Application of nanotechnology in the agricultural and food processing industries: A review. *Sustainable Materials and Technologies* 39, e00809.

Sharifi, S., Mahmoud, N.N., Voke, E., Landry, M.P. and Mahmoudi, M. (2022) Importance of standardizing analytical characterization methodology for improved reliability of the nanomedicine literature. *Nano-Micro Letters* 14(1), 172. DOI: 10.1007/s40820-022-00922-5.

Sharma, S. and Dhadly, D.K. (2023) Nano-inspired biosensors and plant diseases: Recent advances and challenges. *Nanoparticles and Plant-Microbe Interactions* 135–162. DOI: 10.1016/B978-0-323-90619-7.00002-3.

Sharma, K. and Shivandu, S.K. (2024) Integrating artificial intelligence and internet of things (IoT) for enhanced crop monitoring and management in precision agriculture. *Sensors International* 5, 100292.

Sharma, P., Kumar, S., Patel, A., Datta, B. and DeLong, R.K. (2021a) Nanomaterials for agricultural and ecological defense applications: Active agents and sensors. *Wiley Interdisciplinary Reviews: Nanomedicine and Nanobiotechnology* 13(5), e1713. DOI: 10.1002/gch2.202200209.

Sharma, P.K., Dorlikar, S., Rawat, P., Malik, V., Vats, N. *et al.* (2021b) Nanotechnology and its application: A review. In: *Nanotechnology in Cancer Management*. Elsevier, pp. 1–33. DOI: 10.1016/B978-0-12-818154-6.00010-X.

Sharma, A., Nagarajan, J., Gopalakrishnan, K., Bodana, V., Singh, A. *et al.* (2023a) Nanotechnology applications and implications in food industry. In: *Nanotechnology Applications for Food Safety and Quality Monitoring*. Academic Press, pp. 171–182. DOI: 10.1016/B978-0-323-85791-8.00016-1.

Sharma, A., Ranjit, R., Kumar, N., Kumar, M. and Giri, B.S. (2023b) Nanoparticles based nanosensors: Principles and their applications in active packaging for food quality and safety detection. *Biochemical Engineering Journal* 193, 108861. DOI: 10.1016/j.bej.2023.108861.

Sharma, H., Kumar, A., Singh, R., Malik, A. and Juyal, R. (2024) Enhancing food security and sustainability: Harnessing nano-omics-based technologies for quality monitoring. In: Rajput, V.D., Singh, A., Ghazaryan, K., Minkina, T.M. and Al-Tawaha, A.R.M. (eds) *Sustainable Agriculture: Nanotechnology, Biotechnology, Management and Food Security*. De Gruyter, Berlin, German, pp. 191–212.

Sharma, C., Kundu, S., Singh, S., Saxena, J., Gautam, S. *et al.* (2025) From concept to shelf: Engineering biopolymer-based food packaging for sustainability. *RSC Sustainability*. DOI: 10.1039/D5SU00483G.

Shavisi, N. (2024) Electrospun fiber mats based on chitosan-carrageenan containing *Malva sylvestris* anthocyanins: Physic-mechanical, thermal, and barrier properties along with application as intelligent food packaging materials. *International Journal of Biological Macromolecules* 266, 131077. DOI: 10.1016/j.ijbiomac.2024.131077.

Shoukat, A., Pitann, B., Zafar, M.M., Farooq, M.A., Haroon, M. *et al.* (2024) Nanotechnology for climate change mitigation: Enhancing plant resilience under stress environments. *Journal of Plant Nutrition and Soil Science* 187(5), 604–620. DOI: 10.1002/jpln.202300295.

Silvestre, C., Duraccio, D. and Cimmino, S. (2011) Food packaging based on polymer nanomaterials. *Progress in Polymer Science* 36(12), 1766–1782. DOI: 10.1016/j.progpolymsci.2011.02.003.

Simbine, E.O., Rodrigues, L.D.C., Lapa-Guimaraes, J., Kamimura, E.S., Corassin, C.H. *et al.* (2019) Application of silver nanoparticles in food packages: A review. *Food Science and Technology* 39, 793–802. DOI: 10.1590/fst.36318.

Singh, K. (2020) Nanosensors for food safety and environmental monitoring. In: *Nanotechnology for Food, Agriculture, and Environment*. Springer International Publishing, Cham, pp. 63–84. DOI: 10.1007/978-3-030-31938-0_4.

Singh, P., Kim, Y.J., Zhang, D. and Yang, D.C. (2016) Biological synthesis of nanoparticles from plants and microorganisms. *Trends in Biotechnology* 7(7), 588–599. DOI: 10.1016/j.tibtech.2016.02.006.

Singh, T., Shukla, S., Kumar, P., Wahla, V., Bajpai, V.K. *et al.* (2017) Application of nanotechnology in food science: Perception and overview. *Frontiers in Microbiology* 8, 1501. DOI: 10.3389/fmicb.2017.01501.

Singh, R., Mann, B., Sharma, R. and Singh, S. (2019) Application of nanotechnology in functional foods. *Nanoscience for Sustainable Agriculture* 547–579. DOI: 10.1007/978-3-319-97852-9_21.

Singh, P.M., Tiwari, A., Maity, D. and Saha, S. (2022) Recent progress of nanomaterials in sustainable agricultural applications. *Journal of Materials Science* 57(24), 10836–10862. DOI: 10.1007/s10853-022-07259-9.

Singh, R., Dutt, S., Sharma, P., Sundramoorthy, A.K., Dubey, A. *et al.* (2023a) Future of nanotechnology in food industry: Challenges in processing, packaging, and food safety. *Global Challenges* 7(4), 2200209. DOI: 10.1002/gch2.202200209.

Singh, V., Tiwari, R. and Pandey, P. (2023b) Standardization challenges and future prospects of nano-biosensors in food industries. *Food Bioscience* 55, 102845. DOI: 10.3390/jox15030063.

Singh, S., Pathak, S., Singh, J., Kamboj, M. and Singh, V. (2024) Nano-seed coating technologies for enhancing vegetable seed performance and stress tolerance. *Journal of Agriculture and Ecology Research International* 25(6), 278–295.

Singh, P., Pandey, V.K., Sahu, M., Singh, R. and Joshi, R. (2025a) Microbial biosensors for rapid and accurate food quality monitoring: Detection of contaminants, pathogens, and spoilage indicators. *Journal of Food Safety* 45(5), e70032.

Singh, P., Singh, A.K., Singh, V.K. and Kumar, V. (eds) (2025b) *Nanotechnology-based Sustainable Agriculture*. Wiley-VCH. DOI: 10.1002/9783527849949.ch13.

Sirelkhatim, A., Mahmud, S., Seeni, A., Kaus, N.H.M., Ann, L.C. *et al.* (2015) Review on zinc oxide nanoparticles: Antibacterial activity and toxicity mechanism. *Nano-Micro Letters* 7(3), 219–242.

Smaoui, S., Chérif, I., Hlima, H.B., Khan, M.U., Rebezov, M. *et al.* (2023) Zinc oxide nanoparticles in meat packaging: A systematic review of recent literature. *Food Packaging and Shelf Life* 36, 101045. DOI: 10.1016/j.fpsl.2023.101045.

Song, S.S., Xia, B.Y., Chen, J., Yang, J., Shen, X. *et al.* (2014) Two-dimensional TiO2 nanosheets: *In vivo* toxicity investigation. *RSC Adv* 4(80), 42598–42603. DOI: 10.1039/c4ra05953k.

Song, D., Li, G., Liu, Q., Huang, X., Wang, W. *et al.* (2025) Multi-effect coupling enhanced PtPdRhFeCu HEA/N-Cu-ZnSe@ C biosensing device for pesticide residue detection in fruits and vegetables with ultra-low detection limit. *Food Chemistry* 468, 142468. DOI: 10.1016/j.foodchem.2024.142468.

Sridhar, K., Inbaraj, B.S. and Chen, B.H. (2021) Recent advances in nanoparticle-based strategies for improving carotenoid stability and biological activity. *Antioxidants* 10(5), 713. DOI: 10.3390/antiox10050713.

Srivastava, A.K., Dev, A. and Karmakar, S. (2018) Nanosensors and nanobiosensors in food and agriculture. *Environmental Chemistry Letters* 16, 161–182. DOI: 10.1007/s10311-017-0674-7.

Sun, B., Luo, Y., Yang, D., Yang, J., Zhao, Y. *et al.* (2023) Coordinative management of soil resources and agricultural farmland environment for food security and sustainable development in China. *International Journal of Environmental Research and Public Health* 20(4), 3233. DOI: 10.3390/ijerph20043233.

Suresh, V., Arora, K. and Gupta, S. (2023) Precision food processing enabled by sensor-based automation. *Food Bioscience* 54, 102882. DOI: 10.1016/j.fbio.2023.102882.

Takeuchi, M.T., Kojima, M. and Luetzow, M. (2014) State of the art on the initiatives and activities relevant to risk assessment and risk management of nanotechnologies in the food and agriculture sectors. *Food Research International* 64, 976–981. DOI: 10.1016/j.foodres.2014.03.022.

Tang, C.H., Chen, H.L. and Dong, J.R. (2023) Solid lipid nanoparticles (SLNs) and nanostructured lipid carriers (NLCs) as food-grade nanovehicles for hydrophobic nutraceuticals or bioactives. *Applied Sciences* 13(3), 1726.

Tavana, F., Hematian Sourki, A. and Golmakani, M.T. (2024) Enhancing the accuracy, speed, and efficiency of kafirin-PEO electrospun bio-nanocomposite pH indicators with red beetroot extract using image processing. *Food Science & Nutrition* 12(4), 2874–2885. DOI: 10.1002/fsn3.3968.

Tchonkouang, R.D., Lima, A.R., Quintino, A.C., Cristofoli, N.L. and Vieira, M.C. (2023) UV-C light: A promising preservation technology for vegetable-based nonsolid food products. *Foods* 12(17), 3227. DOI: 10.3390/foods12173227.

Tehri, N., Sangwan, S., Dhewa, T. and Puniya, A.K. (2025) Nanotechnology: A boon for sustainable development of functional foods. DOI: 10.1039/9781837673292-00016.

The Business Research Company (2025) Nanotechnology services global market report 2025. Available at: https://www.thebusinessresearchcompany.com/report/nanotechnology-services-global-m market-report (accessed 20 May 2026).

The World Counts (n.d.) Food waste statistics. Available at: https://www.theworldcounts.com/challenges/people-and-poverty/hunger-and-obesity/food-waste-statistics (accessed 20 May 2026).

Thu, H.E., Haider, M., Khan, S., Sohail, M. and Hussain, Z. (2023) Nanotoxicity induced by nanomaterials: A review of factors affecting nanotoxicity and possible adaptations. *OpenNano* 14, 100190. DOI: 10.1016/j.onano.2023.100190.

Tiede, K., Boxall, A.B., Tear, S.P., Lewis, J., David, H. *et al.* (2008) Detection and characterization of engineered nanoparticles in food and the environment. *Food Additives and Contaminants* 25(7), 795–821.

Trotta, F., Silva, S., Massironi, A., Mirpoor, S.F., Lignou, S. *et al.* (2023) Silver bionanocomposites as active food packaging: Recent advances & future trends tackling the food waste crisis. *Polymers* 15(21), 4243. DOI: 10.3390/polym15214243.

Tschiche, H.R., Bierkandt, F.S., Creutzenberg, O., Fessard, V., Franz, R. *et al.* (2022) Analytical and toxicological aspects of nanomaterials in different product groups: Challenges and opportunities. *Nano-impact* 28, 100416. DOI: 10.1016/j.impact.2022.100416.

Tunc, I. and Akyuz, G.A. (2025) Nanosensors from logistics and supply chain perspective: A literature review. *Current Nanomaterials* 10(2), 167–181. DOI: 10.2174/0124054615280118240125065504.

Tutul, M.J.I., Alam, M. and Wadud, M.A.H. (2023) June. Innovative food monitoring system based on IoT and machine learning. In: *2023 International Conference on Next-Generation Computing, IoT and Machine Learning (NCIM)*, IEEE, pp. 1–6. DOI: 10.1109/NCIM59001.2023.10212608.

U.N. Environment (2024) Food Waste Index Report 2024 | UNEP - UN Environment Programme. Available at: https://www.unep.org/resources/publi cation/food-waste-index-report-2024 (accessed 5 December 2024).

Usmani, S.M., Bremer-Hoffmann, S., Cheyns, K., Cubadda, F., Dumit, V.I. *et al.* (2024) Review of new approach methodologies for application in risk assessment of nanoparticles in the food and feed sector: Status and challenges. *EFSA Supporting Publications* 21(9), 8826E. DOI: 10.2903/sp.efsa.2024.EN-8826.

Vagena, I.A., Gatou, M.A., Theocharous, G., Pantelis, P., Gazouli, M. *et al.* (2024) Functionalized ZnO-based nanocomposites for diverse biological applications: Current trends and future perspectives. *Nanomaterials* 14(5), 397.

Vaidya, S., Deng, C., Wang, Y., Zuverza-Mena, N., Dimkpa, C. *et al.* (2024) Nanotechnology in agriculture: A solution to global food insecurity in a changing climate. *Nano-impact* 34, 100502. DOI: 10.1016/j.impact.2024.100502.

Valenzuela-Amaro, H.M., Aguayo-Acosta, A., Meléndez-Sánchez, E.R., Rosa, O., Vázquez-Ortega, P.G. *et al.* (2023) Emerging applications of nanobiosensors in pathogen detection in water and food. *Biosensors* 13(10), 922. DOI: 10.3390/bios13100922.

Vanaraj, R., IP, B., Mayakrishnan, G., Kim, I.S. and Kim, S.C. (2025) A systematic review of the applications of electronic nose and electronic tongue in food quality assessment and safety. *Chemosensors* 13(5), 161. DOI: 10.3390/chemosensors13050161.

Vasile, C., Tantaru, G. and Creteanu, A. (2025) Recent insights into the research of (bio)active additives for advanced polymer materials. *Polymers* 17(23), 3139. DOI: 10.3390/polym17233139.

Venkatesan, R. and Rajeswari, N. (2017) ZnO/PBAT nanocomposite films: Investigation on the mechanical and biological activity for food packaging. *Polymers for Advanced Technologies* 28(1), 20–27.

Verleysen, E., Waegeneers, N., De Vos, S., Brassinne, F., Ledecq, M. *et al.* (2021) Physicochemical characterization of nanoparticles in food additives in the context of risk identification. *EFSA Supporting Publications* 18(6), 6678E. DOI: 10.2903/sp.efsa.2021.EN-6678.

Vieira, I.R.S., de Carvalho, A.P.A.D. and Conte-Junior, C.A. (2022) Recent advances in biobased and biodegradable polymer nanocomposites, nanoparticles, and natural antioxidants for antibacterial and antioxidant food packaging applications. *Comprehensive Reviews in Food Science and Food Safety* 21(4), 3673–3716. DOI: 10.1111/1541-4337.12990.

Villamayor, N., Villaseñor, M.J. and Ríos, Á. (2023) Monitoring nanomaterials in food: A critical overview, perspectives, and challenges. *Exploration of Foods and Foodomics* 1(1), 43–61. DOI: 10.37349/eff.2023.00005.

Visciano, P. and Schirone, M. (2020) Rapid methods for assessing food safety and quality. *Foods* 9(4), 533.

Wang, T., Huang, D., Yang, Z., Xu, S., He, G. *et al.* (2016) A review on graphene-based gas/vapor sensors with unique properties and potential applications. *Nano-Micro Letters* 8(2), 95–119. DOI: 10.1007/s40820-015-0073-1.

Wang, X., Chang, C.H., Jiang, J., Liu, Q., Liao, Y.P. *et al.* (2019) The crystallinity and aspect ratio of cellulose nanomaterials determine their pro-inflammatory and immune adjuvant effects *in vitro* and *in vivo*. *Small* 15(42), 1901642. DOI: 10.1002/smll.201901642.

Wang, M., Li, S., Chen, Z., Zhu, J., Hao, W. *et al.* (2021a) Safety assessment of nanoparticles in food: Current status and prospective. *Nano Today* 39, 101169. DOI: 10.1016/j.nantod.2021.101169.

Wang, X., Liu, Y. and Chen, Z. (2021b) Nano-enabled biosensors for detection of food hazards: Current trends and future perspectives. *TrAC Trends in Analytical Chemistry* 144, 116428. DOI: 10.1016/j.trac.2021.116428.

Wang, C., Ye, M., Zhang, X., Chai, X., Yu, H. *et al.* (2025) Aptamer-based biosensors for rapid detection and early warning of food contaminants: From selection to field applications. *Molecules* 30(22), 4332. DOI: 10.3390/molecules30224332.

Wilson, A.D. (2018) Application of electronic-nose technologies and VOC-biomarkers for the noninvasive early diagnosis of gastrointestinal diseases. *Sensors* 18(8), 2613. DOI: 10.3390/s18082613.

Wójcik, M., Marszałek, K. and Juszczuk-Kubiak, E. (2025) Advances in food processing techniques for allergenicity reduction and allergen identification. *Foods* 14(22), 3933.

Wu, M.T., Zhou, X.G., Chen, M.Y., Wu, A.G. *et al* (2026) Performance-Safety-Feasibility Evidence Chains for Next-Generation Food-Preservation Packaging: Engineering and Regulatory Prioritization of Bio-Based, Nano-Enabled, and Intelligent Systems. *Journal of Food Science* 91(5), 71084. DOI: 10.1111/1750-3841.71084.

Xing, Y., Li, W., Wang, Q., Li, X., Xu, Q. *et al.* (2019) Antimicrobial nanoparticles incorporated in edible coatings and films for the preservation of fruits and vegetables. *Molecules* 24(9), 1695. DOI: 10.3390/molecules24091695.

Xu, Y., Wu, Z., Li, A., Chen, N., Rao, J. *et al.* (2024) Nanocellulose composite films in food packaging materials: A review. *Polymers* 16(3), 423.

Xu, X., Zhu, B., Zeng, Y. and Qi, Y. (2025) Advances in nanomaterials-enhanced biosensing technologies for rapid detection of antibiotic residues in milk. *Microchemical Journal* 115687.

Xuan, L., Ju, Z., Skonieczna, M., Zhou, P.K. and Huang, R. (2023) Nanoparticles-induced potential toxicity on human health: Applications, toxicity mechanisms, and evaluation models. *MedComm* 4(4), e327. DOI: 10.1002/mco2.327.

Yadav, M., Singh, G., Devi, J. and Jadeja, R.N. (2025a) Evaluating food waste: Generation and sustainable disposal strategies. In: *Resource Recycling and Management of Food Waste*. Springer, Cham, pp. 33–48. DOI: 10.1007/978-3-031-86688-3_3.

Yadav, R.K., Mishra, N. and Singh, B. (2025b) Nanotechnology in food packaging. In: *Nanotechnology in Food Packaging*. Springer Nature Switzerland, Cham, pp. 63–87. DOI: 10.1007/978-3-031-82399-2_5.

Yang, D., Zheng, H. and Wang, K. (2025) Review of the application of optical nanosensors in food safety detection. *ACS Applied Electronic Materials* 7(19), 8675–8690. DOI: 10.1021/acsaelm.5c00965.

Yarahmadi, A., Dousti, B., Karami-Khorramabadi, M. and Afkhami, H. (2024) Materials based on biodegradable polymers chitosan/gelatin: A review of potential applications. *Frontiers in Bioengineering and Biotechnology* 12, 1397668. DOI: 10.3389/fbioe.2024.1397668.

Yilma, A.N., Sahu, R., Subbarayan, P., Villinger, F., Coats, M.T. *et al.* (2024) PLGA-chitosan encapsulated IL-10 nanoparticles modulate chlamydia inflammation in mice. *International Journal of Nanomedicine* 1287–1301. DOI: 10.2147/IJN.S432970.

Yosri, N., Gao, S., Zhou, R., Wang, C., Zou, X. *et al.* (2024) Innovative quantum dots-based SERS for ultrasensitive reporting of contaminants in food: Fundamental concepts and practical implementations. *Food Chemistry* 142395. DOI: 10.1016/j.foodchem.2024.142395.

Yousefian Rad, E., Homayouni Tabrizi, M., Ardalan, P., Seyedi, S.M.R., Yadamani, S. *et al.* (2020) *Citrus lemon* essential oil nano-emulsion (CLEO-NE), a safe cell-depended apoptosis inducer in human A549 lung cancer cells with anti-angiogenic activity. *Journal of Microencapsulation* 37(5), 394–402. DOI: 10.1080/02652048.2020.1767223.

Zambrano-Zaragoza, M.L., Quintanar-Guerrero, D., Del Real, A., González-Reza, R.M., Cornejo-Villegas, M.A. *et al.* (2020) Effect of nano-edible coating based on beeswax solid lipid nanoparticles on the preservation of strawberries. *Coatings* 10(3), 253. DOI: 10.3390/coatings10030253.

Zhang, X., Guo, Q. and Cui, D. (2009) Recent advances in nanotechnology applied to biosensors. *Sensors* 9(2), 1033–1053.

Zhang, T., Lv, C., Chen, L., Bai, G., Zhao, G. *et al.* (2014) Encapsulation of anthocyanin molecules within a ferritin nanocage increases their stability and cell uptake efficiency. *Food Research International* 62, 183–192. DOI: 10.1016/j.foodres.2014.

Zhao, P., Tian, Y., You, J., Hu, X. and Liu, Y. (2022) Recent advances of calcium carbonate nanoparticles for biomedical applications. *Bioengineering* 9(11), 691. DOI: 10.3390/bioengineering9110691.

Zhu, Z., Liu, W., Ye, S. and Batista, L. (2022) Packaging design for the circular economy: A systematic review. *Sustainable Production and Consumption* 32, 817–832.

Živković, D., Balanović, L., Mitovski, A., Talijan, N.M., Štrbac, N. *et al.* (2014) Nanomaterials environmental risks and recycling: Actual issues. *Recycling and Sustainable Development* 7(1), 1–8. DOI: 10.5937/ror1401001Z.

Zuo, J., Feng, J., Gameiro, M.G., Tian, Y., Liang, J. *et al.* (2022) RFID-based sensing in smart packaging for food applications: A review. *Future Foods* 6, 100198. DOI: 10.1016/j.fufo.2022.100198.

www.ingramcontent.com/pod-product-compliance
Lightning Source LLC
LaVergne TN
LVHW050437260926
848956LV00008B/29